THE GURKHAS

200 Years of Service to the Crown

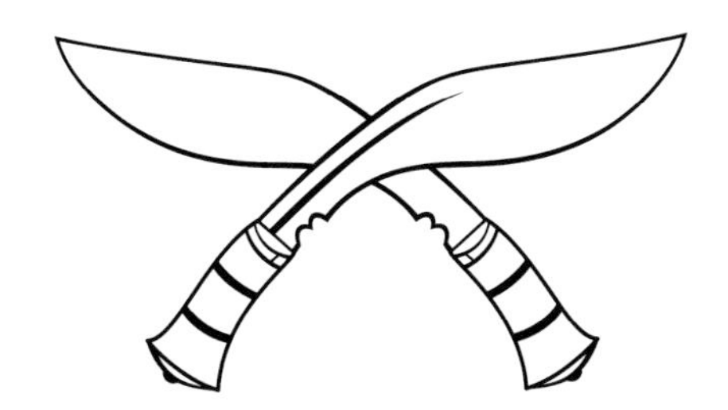

THE GURKHAS

200 Years of Service to the Crown

Major General J C Lawrence CBE

Uniform Press
66 Charlotte Street
London W1T 4QE

Published by *Uniform Press Ltd*, an imprint of Unicorn Publishing Group
www.uniformpress.co.uk

A catalogue record for this book is available from the British Library

ISBN 978 1 910500 02 6

Set in Hiroshige

Design by Nick Newton Design

Printed and bound in the UK by Berforts, Stevenage

Front cover The image on the front cover shows two famous kukris. The one with its handle on the right is the Fisher Kukri which belonged to Major General Frederick Lane Fisher who, as a lieutenant, took part in the Siege of Delhi in 1857 during the Indian Mutiny. The one with its handle on the left belonged to LCpl Tuljung Gurung of the 1st Battalion The Royal Gurkha Rifles who, whilst serving in Afghanistan in 2013, was awarded the Military Cross for his bravery in hand-to-hand combat.

Back cover Soldiers from The Royal Gurkha Rifles practise close quarter fighting with the kukri in Bosnia, 1995

Frontispiece Acting Sergeant Dipprasad Pun of The Royal Gurkha Rifles with his kukri. He was awarded a Conspicuous Gallantry Cross (CGC) for his actions on 17 September 2010 in defence of a remote compound in Helmand Province, Afghanistan. To date, Acting Sergeant Dipprasad is the only Gurkha to have received this award.

Picture Credits:
Page 5 *Major General David Ochterlony* by permission of the National Galleries of Scotland. Page 14 *The Battle of Sobraon* by permission of the National Army Museum. Page 74 *Jallianwalla Bagh, Amritsar* by permission of Mary Evans Picture Library. Pages xii, 8, 32, 52, 72, 98, 140, 164, 202 iStock/Getty images. Photographs on pages 93–96 Crown Copyright and Gurkha Museum, 203–221 Crown Copyright, pages 228–232 by permission of the Gurkha Welfare Trust. All other pictures and photographs by kind permission of Headquarters Brigade of Gurkhas and the Gurkha Museum.

Contents

CLARENCE HOUSE

In April 1815, the Honorable East India Company began recruiting soldiers from the mountain Kingdom of Nepal. Since then, Gurkhas have continued to serve in the British Army with distinction, loyalty and courage. This book tells their remarkable story. It begins in 1814 when the British were halted in their tracks by a small army of mountain warriors and finishes with a remarkable chapter on the contribution that today's Gurkhas have been making in Afghanistan. The skirmishes and battles of the intervening years, including the two World Wars in which fifteen thousand Gurkhas died in the service of the Crown, are illustrated in the two hundred images in this book; one for every year of service. Although the book commemorates and celebrates two centuries of loyal Gurkha service to the Crown, it also aims to raise funds for the work of the Gurkha Welfare Trust; a charity of which I am proud to be Patron. It exists to enable retired Gurkhas to live out their lives with dignity; providing welfare services – ranging from pensions and residential homes for ex-Gurkhas, through to schools and water projects in the most remote regions of Nepal. It is a worthy cause. As the pictures in the book show, the standards, traditions and spirit of the Gurkhas have given a great deal to our country, and they have stood by us in our times of need. I commend both this book and the charity to you in this notable two hundredth year of loyal and dedicated Gurkha service.

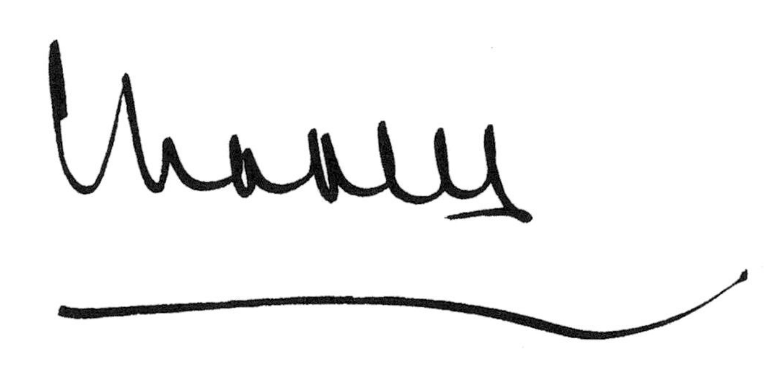

Introduction

Joanna Lumley OBE
Vice Patron of the Gurkha Welfare Trust

My father was a career officer in the 6th Gurkha Rifles and Gurkhas have therefore been a part of my life for as long as I can remember. My parents were based in India when I was born and my early years were spent surrounded by Gurkhas and their families. They are a remarkable people. Proud yet humble, brave yet compassionate, my affection and respect for them runs deep. This is why I felt moved to help them in their quest for fair treatment. It's also why I am proud to introduce this book which tries to convey a sense of what makes them so special. I think it succeeds.

The book chronicles the remarkable service that Gurkhas have provided to the Crown over the last two hundred years. It's a fascinating story and it's remarkable to think that it all started on a lonely hilltop in northern India over two centuries ago. Since then, the Gurkha name has become synonymous with bravery, loyalty and courage. The Gurkha reputation has been hard won. Gurkhas have fought in nearly every conflict that Britain has been involved in for the last two centuries. From the Indian Mutiny and skirmishes on the North West Frontier, through two World Wars and more recent conflicts in the Falklands, Iraq and Afghanistan, Britain's Gurkhas have been in the thick of the fighting, earning 26 Victoria Crosses.

What might surprise many is that Gurkhas are still an important part of the British Army. Recruiting remains vibrant and a place in Britain's Brigade of Gurkhas is as fiercely contested today as it has always been – indeed, there were nearly 8000 applicants for less than 200 places last year! But though Gurkhas are thriving in the modern British Army, many of those who have served over the years are now entering the twilight years of their life.

Joanna Lumley OBE with Gurkhas at a Gurkha Welfare Trust event

The Gurkha Welfare Trust, a charity of which I'm proud to be vice patron, was set up in 1969 when it was apparent that many retired Gurkhas were living in conditions of abject poverty. A large number of these were veterans of the two World Wars. It's a little known fact but Britain recruited hundreds of thousands of Gurkhas during these two major conflicts. As this book shows, they served with distinction but, when peace eventually came, many were returned to civilian life without pensions. As they grew older, their ability to look after themselves in Nepal's harsh environment declined and many found themselves living in conditions of real privation. Every month, the Gurkha Welfare Trust still pays 6667 of these veterans, or their widows, a pension of 8000 rupees. It's not a huge amount of money but it is sufficient to enable them to live out the remainder of their lives with dignity – it is no less than they deserve for the sacrifices they made when Britain most needed them.

But the Gurkha Welfare Trust does far more than just pay pensions. It runs two residential homes for the most frail and provides one-off hardship grants for when our ex-soldiers face seemingly insurmountable problems. Over the years, the Trust's rural water and sanitation programme has installed water and sanitation schemes in 1406 hill villages, benefiting over a quarter of a million people. It has constructed 127 schools and repaired or refurbished another 1544 schools, bringing the gift of education to the families and communities of ex-Gurkhas in the remotest parts of Nepal. Recognising that more and more Gurkhas are now settling in the UK, the Trust has set up two Welfare Advice

Centres in the south of England. These help ex-Gurkhas and their families integrate into UK communities, providing a friendly and sympathetic source of advice. They also ensure that ex-Gurkhas who fall on hard times in the UK are looked after.

I have visited Nepal a number of times. One thing that always strikes me is the absence of decent medical care as you trek away from the urban centres. Life really is hard in the hills. Many families still exist by subsistence farming; it is in many ways a medieval existence. Illnesses that would be easily cured here in the UK kill thousands of people every year. That is why I am extremely proud of the Trust's Mobile Medical Camps. We run eight of these a year. They go up into the mountain villages and provide basic medical, dental, gynecological and ophthalmic care to over 14,500 people every year. There is huge potential to do a great deal more but, to do this, we need more money. It is money well spent and in this 200th year of Gurkha service to the Crown we have decided that this is what we should focus on. Every book that we sell will help provide more mobile doctors and more district nurses to look after those who need our help most. It is a worthy cause and I commend it to you. As you can see from the magnificent images in this book, the Gurkhas have done their bit for us when we have been at our most vulnerable as a nation; it is our turn now to do something for the brave warriors who, in their declining years, are at their most vulnerable.

Joanna Lumley is the Vice-Patron of the Gurkha Welfare Trust. She is shown here visiting ex-Gurkhas who are being helped by the Trust

Author's Preface

The author in Afghanistan towards the end of an 11 month deployment in support of the 2014 presidential elections

For just over two months in 1982, I, like the rest of the Nation, was transfixed as the British Armed Forces fought to re-take the Falkland Islands, a British overseas territory deep in the South Atlantic. I was set on a career in the Army at the time and I was therefore glued to the television as, hilltop by hilltop, British and Gurkha soldiers gradually regained control of the islands. I admired the courage and determination of all those who took part but I was particularly intrigued by the smiling faces of the men of the 7th Gurkha Rifles (7GR). A year later, I went on a potential officers' familiarisation visit to 7GR in Church Crookham, the UK Gurkha Battalion's old base in Hampshire. I had a memorable three days. I had been on numerous visits to other regiments but this was different. Not only was it the most physically and mentally demanding but it was also the most enjoyable. The soldiers impressed from the outset and I came away more determined than ever to join the Gurkhas.

In August 1984, I achieved my ambition when I was formally commissioned into the First Battalion of the 2nd King Edward VII's Own Gurkha Rifles (The Sirmoor Rifles). I have served with Gurkhas ever since. My deep affection and respect for these gentle but capable warriors has continued to grow over the 30 years that I have worked alongside them, both on operations and in peacetime. Brigadier Christopher Bullock, a man who knows more about Gurkhas than I will ever know, produced the definitive history of the Brigade in 2009. This book is not intended to compete with his remarkable work. Rather, I have produced this book because, as we celebrate and commemorate 200 years of Gurkha service to the Crown, I wanted a way of explaining what, to me at least, makes this such a special achievement. I hope that the narrative and images in this book help people to understand why the 200th anniversary of Gurkha service is worth celebrating and why Sir Ralph Turner MC felt moved to describe the Gurkhas as *'bravest of the brave, most generous of the generous, never had a country more faithful friends than you.'*

The author, as colonel of The Royal Gurkha Rifles, on a visit to the Second Battalion in March 2015

CHAPTER 1

The British Empire Meets the Gurkhas

1814–1816

On 31 October 1814, Major General Sir Rollo Gillespie, an experienced and brave field commander,[1] was killed leading a force of East India Company and British troops against a remote hill fort in Kalunga, a small town in northern India. Despite being equipped with the most modern weaponry of the time, it took the British nearly a month to capture the fort. The defending force, which comprised only 650 soldiers from Nepal, inflicted heavy casualties on Gillespie's force of some 4,000 troops. That such a small contingent of 'native' troops had been able to halt the advance of a vastly superior force from one of the most competent armies in the world was a remarkable achievement. It shocked British India.[2] Writing in 1819, the historian Henry T Prinsep, then an officer of the Honourable East India Company's Civil Service, noted that '… the sinister shadow of these events, in damping the ardour of our own troops, and in giving courage to those of the enemy and hopes to the malcontents in every part of the ample surface of India, was for a long time counteracted by no one brilliant exploit in our arms'.[3]

Major General Sir Rollo Gillespie who was killed at the Battle of Kalunga on 31 October 1814 leading British and East India Company troops against the Gurkhas of Balbahadur Kunwar

The 'natives' at Kalunga (or Nalapani as it is now known) were Gurkhas, fighting men from the mountains of Nepal. Disciplined, well trained and commanded by capable and experienced officers, the Gurkhas had seized and then occupied vast tracts of India, then comprised of a multitude of petty principalities,[4] to the west, south and east of Nepal's own borders. With its own interests in northern India threatened by Nepal's enthusiastic expansion, the Supreme Government of the Honourable East India Company had dispatched an army of 30,000 troops, 60 guns, 12,000 Indian auxiliaries, 1113 elephants and 3682 camels[5] to bring the 'aggressive little state of Nepal'[6] to heel. The army was organised into four fighting columns. Two were sent to the west, one of which was Gillespie's, to regain territory seized by the Nepalese. The other two columns were dispatched towards the centre of Nepal and its capital, Kathmandu. The Battle of Kalunga was the first significant

The monuments at Kalunga. As a mark of respect for their 'gallant adversary', the British erected two memorials, one to their own fallen and one to the Gurkhas who had opposed them so courageously (Gurkha Museum)

The inscription on the monument to the Gurkhas at Kalunga (Gurkha Museum)

Bhim Sen Thapa who was prime minister of Nepal at the time of the Battle of Kalunga

engagement of the war. Although the Gurkhas were eventually defeated at Kalunga, their bravery and soldierly competence made an immediate impression on the British to the extent that they erected two monuments at Kalunga, one to their own fallen and one to the fallen of their 'gallant adversary'.[7]

At the time of the Battle of Kalunga, Nepal was controlled by its prime minister, Bhim Sen Thapa. The monarchy, though it existed, had been marginalised in 1806 when the then king, Ranbahadur, was killed by his half-brother, leaving a childless queen and the infant son of one of Ranbahadur's mistresses as the heir apparent.[8] An unpleasant and cruel man, Ranbahadur came from more impressive stock. His grandfather, Prithvi Narayan Shah, had been the ruler of a small mountain kingdom in the west of Nepal called Gorkha. Ruthless and ambitious, Prithvi Narayan had expanded his kingdom until, by around 1768, he had succeeded in unifying Nepal.[9] Prithvi Narayan's men were known as Gorkhas, or Gurkhas, because they came from the Kingdom of Gorkha. Though Prithvi Narayan was long dead and the country was controlled from Kathmandu, those opposing the British at Kalunga retained the Gurkha sobriquet because they fought for the reigning House of Gorkha.

Gillespie's defeat was a blow to British morale but worse was to come as the British sought to impose their will on the Nepalese. On 27 December 1814, Major General Martindell, who had taken over from the unfortunate Gillespie, tried to take the hill fortress of Jaithak in the far west of Nepal's captured territories. Repulsed on one front, he failed to exploit success on another and withdrew with '… the loss of many valuable lives'.[10] The triumphant Gurkhas hailed his retreat as a significant victory.[11] The column commanded by Major General Bennet Marley was ambushed on New Year's Day 1815 as it tried to make its way through the low country to the south of Nepal's capital, causing its commander to withdraw and lose any appetite he might have had for further offensive operations.[12] The column commanded by Major General John Wood was similarly deterred from fighting through into the mountains of Nepal after two of its reconnaissance parties were engaged by Gurkhas.[13] Only Major General David Ochterlony, an astute

Gurkhas circa 1815. The picture was commissioned by William Fraser whilst the Deputy Resident of Delhi (Gurkha Museum)

and patient man who commanded the remaining column in the west,[14] achieved success. This is the more impressive because he found himself facing the main Gurkha army in well prepared defensive positions. Comprised of about 3000 troops, the Gurkha army was commanded by General Amar Singh Thapa, one of Nepal's most competent generals. Ochterlony's force, which numbered about 6000 men,[15] attacked and, over the course of several weeks of hard fighting, succeeded in driving Amar Singh back, ridge by ridge,[16] to the heavily defended mountain fortress of Malaun.[17]

Eventually, on 15 April 1815, Ochterlony managed to out-manoeuvre Amar Singh by getting two of his cannons onto the high ground that dominated the fortress at Malaun. Amar Sing mounted a spirited counter-attack but failed to dislodge Ochterlony's force and, in particular, his lethal artillery. Realising that his position had now become untenable, Amar Singh, whose force had been reduced to about 200 men, was left with little option but to seek a settlement. On 15 May 1815, Amar Singh marched out of Malaun, ceding the fortress to the British, and headed back towards Nepal to join up with his son, Ranjur Singh, who had been defending the mountain fortress at Jaithak. As a mark of respect for his adversary, Ochterlony allowed the Gurkhas to take their personal property and weapons with them as they withdrew.

The defeat of Amar Singh's army had come at a significant cost, both in terms of life and of reputation. The Supreme Government of the East India Company had been forced to field its biggest ever army and it was in no mood to be lenient with the Nepalese.[18,19] They drafted a peace treaty, known as the Treaty of Segauli, which imposed tough conditions.[20] Not only was Nepal required to relinquish the territories it had seized in the west and east but it also had to give up large tracts of its own most fertile territory, known as the Terai, in the south. But senior members of the ruling elite owned considerable amounts of land in the Terai and, as it formed a large part of their wealth, they had no intention of giving it up easily.[21] Negotiations continued throughout the remainder of year, with the Treaty being re-drafted a number of times. By the end of December 1815, it was apparent that the Nepalese had no intention of ratifying it.[22]

Encouraged by General Amar Singh Thapa to believe they could defend their heartland against a British attack, the Nepalese started to prepare defensive positions in the mountains to the south of Kathmandu. At the same time, they continued 'hollow' discussions with the British,[23] hoping to delay the onset of war until the rainy season arrived. But the Nepalese had miscalculated and, in January 1816, an army of nearly 17,000 men under the command of Major General David Ochterlony deployed onto the Terai to begin its advance on Kathmandu.[24,25] At first, Ochterlony encountered little opposition, crossing the Terai in short order and entering the foothills to the south of the Gurkhas' main defensive position at Makwanpur. He followed the line of a valley but his route to Makwanpur was blocked by a well prepared defensive position in a narrow mountain pass known as Chereea-ghatee.[26] Had he tried to take the position by frontal assault, he would have sustained a high number of casualties.[27] However, Ochterlony had learned of an old smugglers' path which would bring him out behind the Gurkhas' positions. Showing considerable personal courage, he led his men along the path, enduring 'great toil and privations' for two days, until he was in a position to turn the Gurkhas' defences.[28,29] Realising their predicament, the Gurkhas withdrew back up

Major General David Ochterlony, the talented field commander who eventually defeated the Gurkhas in 1816 at the Battle of Makwanpur

the valley towards the defended fortress at Makwanpur.[30] Ochterlony rejoined the rest of his force, which included 83 cannons, and pressed forward, attacking the Gurkhas' positions. The Gurkhas fought with great courage but they were outnumbered and outgunned. John Shipp, then an Ensign in the 87th Foot, observed that 'I have never seen more steadiness or bravery exhibited in my life. Run they would not and of death they seemed to have no fear, though their comrades were falling thick around them, for we were so near that every shot told'.[31] After a particularly bloody day of fighting on 28 February 1816, during which they lost 800 of their 3000 soldiers, the Gurkhas realised that they could not win against such an overwhelming force.[32] On 4 March 1816, the Nepalese signed the Treaty of Segauli, bringing Britain's war with Nepal to an end.[33]

From the outset of the campaign, the Gurkhas had impressed Ochterlony with their fighting qualities. Tough, disciplined and brave, he recognised that they had a great deal to offer as soldiers of the East India Company and, having obtained permission from the commander-in-chief, he assembled a force of some 2000 irregulars from the many deserters and prisoners captured by the British. Lieutenant Frederick Young, in whose arms Gillespie had died at the Battle of Kalunga, was appointed as the commander. Although Young did not know it at the time, many of his men came from towns and villages that had been seized by the Gurkhas and not from Nepal itself. Lacking the martial spirit of the true Gurkha, they fled in their first engagement, leaving Young to be taken prisoner. His captors apparently asked him why he had not run away like his men. His response was that he hadn't come all the way from England to run away.[34] Impressed, the Gurkhas replied that they could serve under men like him!

Frederick Young who commanded the first Gurkha unit to fight for the Honorable East India Company and raised the Sirmoor Battalion (later to become the 2nd Gurkhas) in 1815

In April 1815, the East India Company formally authorised the recruitment of four battalions of Gurkhas.[35] The first of these were the two Nusseree (or 'friendly') Battalions raised by Lieutenant Ross which would later amalgamate and then become the 1st Gurkha Rifles. Frederick Young, now released from captivity, raised a further unit, the Sirmoor Battalion, which later became the Sirmoor Rifles and then 2nd Gurkha Rifles. The last unit to be raised in 1815 was the Kumaon Battalion, later to become the 3rd Gurkha Rifles.[36] Interestingly, one of the Nusseree Battalions, under the command of Lieutenant Lawtie, saw service against Amar Singh at Malaun;

A Gurkha soldier circa 1815. As well as carrying a kukri and a sword, they were also equipped with muskets

Lawtie afterwards reported that he '... had the greatest reason to be satisfied with their exertions'.[37]

200 years have now passed since the first Gurkhas were recruited to serve the British Crown. That they have remained a valued and relevant part of the British Army for two centuries owes much to their fighting spirit, demonstrated time and time again in the many battles and skirmishes illustrated in the following chapters of this book. There are well over 200 images, more than one for every year of service, and they tell a remarkable story.

The signing of the Treaty of Segauli on 4 March 1816 which brought Britain's wars with Nepal to an end

CHAPTER 2

The Gurkhas Prove their Worth (Time and Again)

1817–1858

The newly formed Gurkha battalions settled into the routine of garrison life in northern India. They established their regimental homes in territory that had previously been annexed by Nepal: the Nusseree Battalion (later 1st Gurkhas) in Dharamsala; the Sirmoor Battalion (later 2nd Gurkhas) in Dehra Dun; and the Kumaon Battalion (later 3rd Gurkhas) in Almorah.[1] Life was relatively quiet until, in October 1824, 350 men of the Sirmoor Battalion, still commanded by Frederick Young, were dispatched to Koonja to suppress a minor uprising. The leader of

Sketch of Gurkha soldiers of the Sirmoor Battalion at Dehra Dun circa 1821. The sketch was done by Frederick Shore, the assistant magistrate at the Sirmoor Battalion's home base of Dehra Dun

the insurgents (or 'dacoits' as they were known) had styled himself as the Rajah Kullian Singh and had sent out notices to local landholders demanding payment in exchange for not plundering their villages.[2] The dacoit gang numbered between eight and nine hundred and had taken refuge in a small fort near the village of Koonja.[3] The Sirmoor Battalion marched a remarkable 27 miles over rough country in a single day and arrived at the fort on 3 October 1824.[4]

The fort was in good repair and the dacoits felt secure enough to taunt the Gurkhas, calling them 'hill monkeys' and threatening to cut them up.[5] However, their fortunes were about to change. Mr F J Shore, the joint magistrate and assistant commissioner from the Sirmoor Battalion's home base of Dehra Dun, had travelled with the Gurkhas and suggested chopping down a young tree to use as a battering ram against the fort's main gate.[6] Enthused by the idea, the Gurkhas did as suggested

A sketch showing the Sirmoor Battalion using a tree as an improvised battering ram to storm the fort at Koonja. The sketch was done by Frederick Shore, the assistant magistrate in Dehra Dun, who had accompanied the Sirmoor Battalion to Koonja and whose idea it was to use a young tree as a battering ram

and, within a short period of time, had smashed their way into the fort. The fighting was intense. As Shore notes, the dacoits '... fought sword in hand to the last with that desperation, which the certainty of being killed on one side or hanged on the other, generally inspires.'[7] Forty of the Gurkhas were killed or wounded but the dacoits fared far worse, losing nearly a third of their number. The Gurkhas' victory was important. Unrest was brewing in northern India, fuelled by a rumour that the British were going to evacuate the north-western provinces, and as Shore notes '... the effect of the storm of this fort was, to use the words of a neighbouring magistrate, like an electric shock: it happened at an auspicious moment and was productive of important benefit.'[8] Interestingly, the memory of the Battle of Koonja lives on in the contemporary Brigade of Gurkhas; to this day, officers of The Royal Gurkha Rifles wear a silver ram's head on their cross-belts to commemorate the use of the improvised Koonja battering ram.[9]

Baldeo Singh, the raja of Bhurtpore, was one of the rulers of the many petty principalities that characterised India at the time. A shrewd man, in 1824 he cut a deal with the East India Company's then political agent in Delhi (David Ochterlony) that, in the event of his death, the British would support this son, Bulwant Singh, as his successor.[10] Regrettably, the old raja died '... not without suspicion of poisoning' whilst on a pilgrimage to Goverdhan,[11] then, as now, a town of great religious significance to Hindus. The young raja, then no more than five or six years of age, succeeded his father under the guardianship of his maternal uncle, Ram Ratan Singh.[12] Unfortunately, the old raja's nephew, Durjan Sal, killed Ram Ratan Singh and assumed power for himself.[13]

A man of his word, Ochterlony immediately assembled a force to re-impose the succession he had agreed with the old raja. However, the governor general, Lord Amherst, had reservations about Ochterlony's ability to restore order. Concerned that any failure on the part of the British '... might endanger the stability of the British Indian Empire',[14] he ordered the army assembled by Ochterlony to stand down. Ochterlony reluctantly complied but, only a few months later, the Commander-in-Chief, Lord Combermere, had little option but to re-assemble the force as word of the new ruler of Bhurtpore's open defiance started to spread. The force assembled to take Bhurtpore numbered between 21,000 to 27,000, including 100 men from each of the Nusseree and Sirmoor Battalions.[15,16] Bhurtpore was considered to be the most impregnable fortress in India – indeed, the Indians believed '... that it could never be taken by

The Siege of Bhurtpore. The picture shows the blowing of a huge mine under the fort's ramparts on 26 January 1826 which breached the walls, allowing British and East India Company troops to storm the fort

mortal man.'[17] The army dispatched to lay siege to Bhurtpore therefore included every artillery piece in northern India.[18] The fortress' walls, which were 60 ft thick, ran for five miles and it fast became apparent that it would take more than artillery to break into the well-defended fortress.[19]

In early January 1826, the British started to mine under the ramparts of the great fortress. The Gurkhas, acting as skirmishers and marksmen, were 'conspicuously distinguished' in the way that they prevented the enemy from interfering with the force's mining operations.[20] On 18 January 1826, the British detonated a 10,000 lb mine that they had succeeding in placing under the walls.[21] The explosion created a breech, allowing British and East India Company troops to surge into the fortress. The two detachments from the Nusseree and Sirmoor Battalions were in the midst of the main assault. The fighting that followed was particularly fierce with approximately 14,000 of Durjan Sal's followers being either killed or wounded; the victor's losses amounted to less than 600.[22] On 20 January 1826, the young raja was restored to the throne of his ancestors.[23] From the British perspective, the Siege of Bhurtpore was another important victory. Not only did it remove an immediate threat to British dominance but it sent a powerful message to others who might also have been contemplating open defiance of their imperial masters.

On 11 December 1845, the Sikh army crossed the River Sutlej, the disputed border between the powerful Sikh kingdom of the Punjab and the East India Company's northwestern territories. Exactly why the Sikhs took such a provocative step remains unclear. One theory is that it was to pre-empt the British. Although the East India Company had signed a non-aggression treaty with the Sikh Durbar in 1809, they had been amassing forces, including pontoons and bridging equipment, on their side of the border for several months before the Sikhs crossed the Sutlej. It is therefore possible that the British had intended to launch an attack of their own, exploiting the disorder that had characterized the state since the death of its long time ruler, Maharajah Ranjit Singh, in 1839 in order to remove the last remaining force capable of challenging British dominance in northern India. Another theory is that it was part of a 'duplicitous plan' devised by one of the two rival factions within the Punjab to weaken the Sikh army which, large and well trained, had become too powerful.[24]

A Gurkha soldier circa 1821

Whatever the reason, when the Sikhs crossed the River Sutlej the East India Company yet again found itself having to suppress a perceived challenge to its supremacy. The Nusseree and Sirmoor Battalions were deployed as part of an Army under the command of Lieutenant General Sir Hugh Gough, the commander-in-chief of the East India Company's Bengal Army. Of the four battles that make up what historians call the First Anglo-Sikh War, the Gurkha battalions were involved in at least two, the Battle of Aliwal on 28 January 1846 and the Battle of Sobraon on 10 February 1846. The Sikhs, trained by Europeans,[25] were capable soldiers and, whilst their senior leadership might have been in disarray, they fought with courage and determination. The Sirmoor Battalion alone lost 49 dead at Aliwal whilst the Nusseree Battalion lost six dead with a further 16 wounded.[26] The casualties increased at the Battle of Sobraon. The Sikh forces, equipped with 70 cannon and numbering about 35,000,[27] had concentrated

The Battle of Sobraon which took place on 10 February 1846. Both the Nusseree and Sirmoor Battalions (later the 1st and 2nd Gurkhas) fought with distinction during this final major engagement of the First Anglo-Sikh War

in a bridge-head on the eastern bank of the Sutlej. They held a fortified position which dominated the approaches to the river and were resupplied by a bridge of boats which crossed the Sutlej to their rear.

Opposing them were 20,000 British and East India Company troops.[28,29] After a lengthy artillery duel, the British assaulted the fortifications and the Gurkha battalions once again found themselves in the thick of the action. Tej Singh, the Sikh commander, fled the battle and retreated back across the Sutlej. Damage to the bridge inflicted by British artillery, and by Tej Singh's men firing the bridge as they escaped, ensured that the main Sikh army was unable to withdraw.[30] Trapped, the Sikhs fought to the death. Though eventually defeated, they inflicted tremendous casualties on the British and East India Company troops, with the total killed and wounded estimated at 2383.[31] Gurkha casualties were particularly heavy. Of the 610 men of the Sirmoor Battalion, 145 were killed or wounded, including the commandant, Captain John Fisher, who died leading his men from the front.[32] The Nusseree Battalion suffered seven killed and 77 wounded.[33]

Captain John Fisher, commandant of the Sirmoor Battalion, killed leading his Gurkhas against the Sikhs at the Battle of Sobraon, 10 February 1846

Gurkhas during the Anglo-Sikh wars. Note the reference to the Gurkhas' famous kukri fighting knife in the caption 'the Nusseree Battalion teaching the Sikhs the art of cookery'

In his dispatches after the Battle of Sobraon, Sir Hugh Gough observed:

> I must pause in this narrative especially to notice the determined hardihood and bravery with which our two Battalions of Ghoorkhas, The Sirmoor and Nusseeree, met the Sikhs, wherever they were opposed to them. Soldiers of small stature but indomitable spirit they vied in ardent courage with The Grenadiers of our own nation, and, armed with the short weapons of their mountains, were a terror to the Sikhs throughout this great combat.[34]

In 1857, the Gurkhas were again called upon to demonstrate not only their fighting skills but also their loyalty as the native troops of British India started to turn against their imperial masters. The violence started on 11 May in the town of Meerut. Although it came as a surprise to the British, it had been brewing for years. Arguably, the main catalyst was the cartridge that accompanied the new Enfield rifle which the firer had to rip open with his teeth. This was usual practice with weapons of the time but rumours soon began to circulate that the cartridge was greased with cow or pig fat. As a sacred animal, the use of cow fat was offensive to Hindus and, to Muslims, the pig was considered unclean. Native soldiers were therefore reluctant to put the cartridges to their mouths. The Gurkhas, ever practical, had few such reservations, even demonstrating

Soldiers of the Sirmoor Battalion (later the 2nd Gurkhas) in front of Hindu Rao's House on Delhi Ridge. The ridge dominated the city of Delhi which, in 1857, became the focus of the Indian Mutiny. Despite suffering tremendous losses, the Sirmoor Battalion, along with the 60th Rifles and the Corps of Guides, succeeded in holding the ridge until reinforcements arrived

to other native troops on a musketry course that the new cartridges were fine.[35] However, discontent turned into open rebellion when native troops were forced to use the cartridges against their will.

On 14 May 1857, the Sirmoor Battalion was ordered to proceed to Meerut to help suppress the rebellion.[36] But events were moving fast and the battalion was soon redirected to Delhi to reinforce a British column under Lieutenant General Sir Henry Barnard.[37] The battalion's arrival was greeted with some suspicion. Worried that the Gurkhas might also mutiny, their tents had been pitched close to the artillery which, unbeknownst to the Gurkhas, had been given orders to turn their guns on the battalion at the first sign of rebellion.[38]

The Gurkhas soon found themselves in the thick of the action, fighting alongside the 60th Rifles, and any doubts about their loyalty were quickly dispelled. The rebels consolidated their position by withdrawing into the heavily fortified city of Delhi. This allowed the British to occupy a ridge to the north east which overlooked the approaches to the city. The mutineers realised the tactical importance of the ridgeline and, in particular, a sprawling villa which dominated the ridge belonging to a local merchant called Hindu Rao. But the men of the 60th Rifles, the Sirmoor Battalion and the Corps of Guides held firm, repulsing 26 separate attacks between the first week of June and the 14 September 1857.[39] Conditions were dire. As well as the constant artillery bombardments and rebel attacks, the defenders had to contend with cholera. The casualties continued to mount but the defenders fought on. Eventually, in mid-August a relief column arrived, headed by Brigadier General John Nicholson. Known as the Lion of the Punjab, he commanded a force of some 4200 men, including 90 reinforcements for the Sirmoor Battalion which had lost well over half of its men.[40]

Bahadur Shah Zafar, the elderly King of Delhi, who became the reluctant figurehead of the Indian Mutiny

On 14 September 1857, the defenders on the ridge, reinforced and determined to put an end to nearly four months of constant combat, went on the offensive. Organised into four fighting columns, their aim was to break into the city in order to take the fight to the mutineers.[41] Brigadier General Nicholson, with the Kumaon Battalion in the vanguard of his force, led his column towards the mighty Kashmir Gate. The mutineers resisted, mortally wounding Nicholson,[42] but eventually the gate was blown and the city's formidable defences breached. The attacking troops surged into the city, driving the mutineers from street to street. The fighting lasted for six days; the attackers showed little mercy. The elderly King of Delhi, who had become the mutineers' *de facto* figurehead, was captured. Although three of his sons were killed, his life was spared, a remarkably lenient act given he had offered a 10 rupee reward for the head of every Gurkha.[43,44]

Whilst the Sirmoor Battalion was fighting for its life on the ridge at Delhi, the Nusseree Battalion, renamed the 66th or Goorkha Regiment, was displaying the same courage and determination over to the east in the Kumaon hills. The regiment was part of a British force which was locked in combat with two rebel groups, each about 1000 strong.[45] Seizing the initiative, the 66th

Soldiers from the 66th or Goorkha Regiment (previously known as the Nusseree Battalion) circa 1857. During the Indian Mutiny, the regiment was involved in heavy fighting with mutineers in the north of India. Lieutenant John Tytler, one of the regiment's British Officers, was awarded the Victoria Cross for his actions, becoming the first recipient of this award in a Gurkha battalion

conducted a night march through a dense forest to surprise the rebels at the village of Charpura.[46] Although outnumbered, the Gurkhas routed the rebel group, inflicting heavy casualties.

The re-taking of Delhi was arguably the turning point of the mutiny but it took well over a year for the last embers of rebellion to be extinguished.[47] One of the most significant skirmishes of this period, at least from the Gurkha perspective, occurred on 10 February 1858. Lieutenant John Adam Tytler, a 33-year-old officer of the 66th, was part of a force which found itself facing two rebel groups of about 5000 infantry and 1000 cavalry near the town of Haldwani.[48] Although heavily outnumbered, the British had little option but to fight. The enemy's artillery was proving to be devastatingly effective until Tytler, alone, rode forward to the guns and started to engage the enemy in hand to hand combat. Shot

A group of officers of the 4th Gurkhas. Seated in their midst is Lieutenant John Tytler of the 66th who was awarded the Victoria Cross for his actions during the Indian Mutiny

in the arm and with a spear in his chest, he continued to fight until the guns had been taken.[49] Perhaps not surprisingly, he was awarded the Victoria Cross, the first Gurkha Officer to receive the award.

One enduring memory of this historic event is the relationship that was formed between the Gurkhas of the Sirmoor Battalion and the men of the 60th Rifles. Both recognised in the other the fighting spirit that defines the professional soldier. In recognition of their courage and soldierly conduct, the Gurkhas of the Sirmoor Battalion were allowed to call themselves Riflemen, rather than sepoys, and were permitted to adopt the scarlet facings of the 60th Rifles.[50] These two reminders of this historic relationship live on; soldiers in today's Royal Gurkha Rifles are still known as riflemen and the scarlet, or 'lali' as the Gurkhas call it, remains a defining feature of their dress uniforms.

The bravery of the Sirmoor Battalion at Delhi also led to the battalion being awarded a third colour. But as rifle regiments did not traditionally carry colours, this was replaced with a commemorative Truncheon, designed personally by Queen Victoria.[51] The Queen's Truncheon, as it is known, is still carried with immense pride by today's Royal Gurkha Rifles. It is accorded the respect of a military colour and, when paraded

The two Colours of the 2nd Gurkhas at the time of the Indian Mutiny. A third Colour was awarded in recognition of the regiment's performance during the Indian Mutiny. When the 2nd Gurkhas became a rifle regiment, the third Colour was replaced by a Truncheon specially designed by Queen Victoria

formally, is carried by a Gurkha officer – the Truncheon Jemedar – and escorted by four armed guards whose drill and turnout are expected to be immaculate. On very special occasions, the Queen's Truncheon is presented to the reigning sovereign for formal inspection. The last time that this happened was in May 2015 as part of the formal commemorations to mark 200 years of Gurkha service to the Crown.

One other enduring link that was forged in the heat of the Mutiny is the special relationship that Britain continues to enjoy with Nepal. In 1857, and aware that the British were struggling to contain the Mutiny, the then prime minister of Nepal, Jangabahadur Rana, not only allowed the British to raise additional Gurkha battalions but also offered to come to the East India Company's aid by dispatching an army of his own to help suppress the rebels.[52] Although initially refused, his offer was eventually accepted and, in December 1857, Jangabahadur Rana marched into India at the head of an army of around 5800 men comprised of six regiments of his own Gurkhas.[53] They joined forces with the army of General Sir Colin Campbell and made a significant contribution in the battle to re-take Lucknow.[54]

The Truncheon Jemedar of the 2nd Gurkhas with a Duffadar of the 1st Bengal Lancers (Skinner's Horse) circa 1897

Taken in the 1970s, the photograph shows young soldiers of the 2nd Gurkhas swearing allegiance to the Crown on the Queen's Truncheon which replaced the regiment's three Colours after the Indian Mutiny. The Queen's Truncheon was personally designed by Queen Victoria and is still in service with today's Royal Gurkha Rifles

The Indian Mutiny had cost Britain a huge amount in blood and treasure. Recognising that the Honourable East India Company had outlived its usefulness, the company was abolished through the Government of India Act on 2 August 1858.[55] The British Government assumed direct control of the country, appointing a viceroy to be the in-country representative and a secretary of state to represent India's interests in London.[56] In recognition of the valuable contribution made by the Nepalese Army, Her Britannic Majesty's Government also returned those parts of the terai which had been seized from the Nepalese at the end of the Anglo-Nepal wars.[57]

The Gurkha Kukri

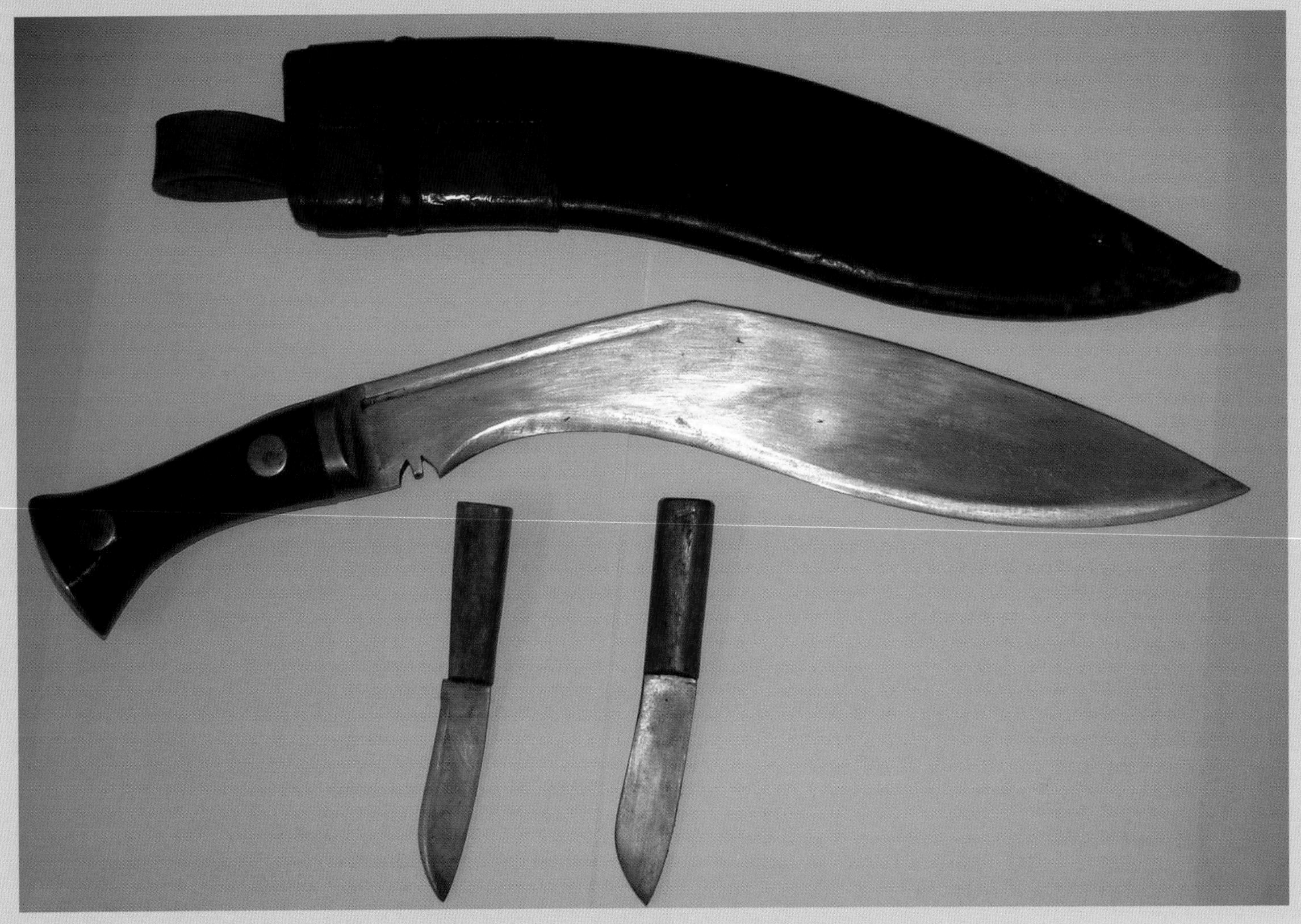

The blade of a modern military kukri is usually about 30cm in length. The scabbard contains two pockets at the back to hold a pair of small knives. One of these, the *chakmak*, is for sharpening the kukri and can be used with a flint to create a spark. The other, the *karda*, is used as a penknife for skinning animals

The Gurkhas are famous for their fighting knife, known as the kukri or 'khukuri'. Short, broad-bladed and with a distinctive curve, it is widely used in the hills of Nepal for chopping wood, killing animals, opening cans, clearing undergrowth and indeed any other task that requires a strong blade. Young men learn to use the kukri from a very early age and it is this deep familiarity with the weapon that makes it so effective in the hands of a Gurkha soldier. By the time a young man joins the Army, the weapon has effectively become an extension of his dominant arm.[1]

There are a number of different theories about the origin of the kukri. One theory is that it is a descendent of the *machaira*, the curved cavalry sword of the ancient

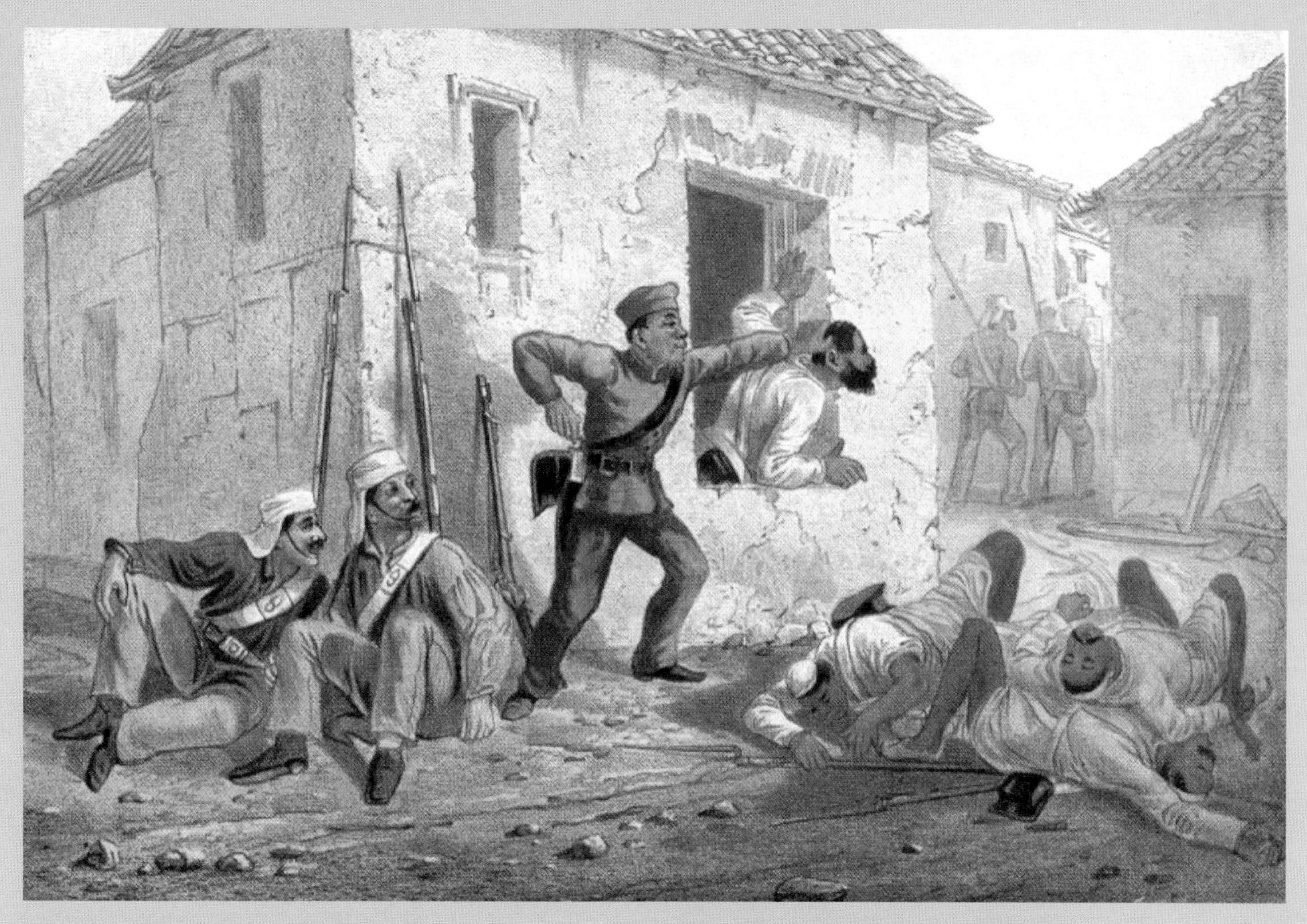

A painting of the storming of Delhi in 1857 showing 'the incident at Subjee Mundi'. A soldier from the Sirmoor Battalion (later the 2nd Gurkhas) is about to draw his kukri and decapitate a mutineer

ancient Macedonians carried by Alexander's horseman when he invaded north west India in the 4th century BC[2]. Another not necessarily contradictory theory is that it originates from a form of knife used by the Mallas who came to power in Nepal in the 13th century.[3] Arguably the most credible theory is that it was developed in isolation by the peasants of Nepal. Its size and dimensions may have been shaped by the environment as a longer weapon would have been impractical given the very steep hillsides that characterise much of Nepal.

Although all kukris have a similar basic shape, there are a number of different variations on the theme.

A rifleman of the 3rd Gurkhas taken in the Regimental home of Almorah circa 1907. The kukri is worn on the soldier's right hip with the handle free of obstruction, enabling the kukri to be easily drawn when required

A group of Gurkhas from 3rd Gurkhas circa 1890. Note the kukris being held by the two soldiers bottom right

A soldier from the 9th Gurkhas demonstrates the use of the kukri to incapacitate an adversary as a group of senior officers approach. Although this photograph was taken in 1945, young Gurkha soldiers are still taught the same slashing action during recruit training

Historically, kukris from the west of Nepal tended to be short and 'round-bellied' whilst those from the eastern districts had longer, more slender blades.[4] Kukris used for ceremonial or sacrificial purposes, such as chopping the head off a water buffalo during the Hindu festival of 'Dashera' or 'Dashain', are necessarily bigger and heavier. *Kothimora* kukris, which are frequently given as gifts to esteemed people, are highly polished and often have intricate silver designs on the scabbard.

The blades of modern military kukris tend to be about 30cm in length and are made of steel. They have a distinctive notch near the handle known as the *kaura*. There are numerous interesting explanations of its presence.[5] One of these is that it is an ingenious aiming sight for when the kukri is thrown at a target. Another is that it is to stop blood running down the blade and onto the handle. Yet another is that it is to catch and then neutralise an enemy blade. Whilst the latter two

A rifleman of the 10th Gurkhas smiles as he tests the edge of his kukri. The photograph was taken in Italy in 1945

Gurkhas from The Royal Gurkha Rifles in Afghanistan

Soldiers from The Royal Gurkha Rifles practising their kukri drills in Bosnia in 2005. Its size and shape make the kukri an ideal weapon for close quarter fighting

Young Nepalese men grow up using the kukri as a general utility weapon in Nepal. During their basic training in the UK, they are taught how to turn this familiarity with the kukri into a lethal capability

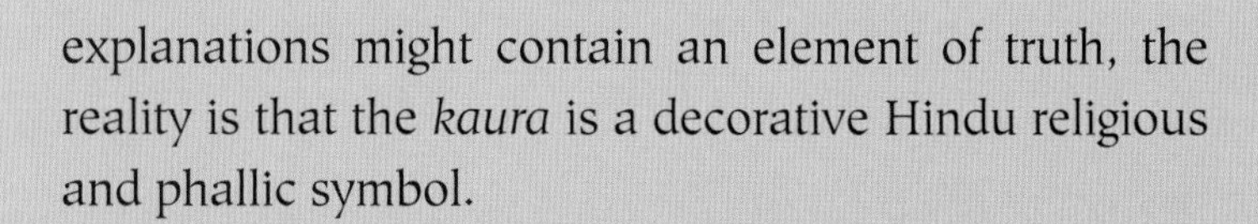

explanations might contain an element of truth, the reality is that the *kaura* is a decorative Hindu religious and phallic symbol.

The handle of the modern military kukri is usually made of horn or dense wood. It is secured to the blade by rivets through the hilt or by flattening the end of the hilt over the bottom of the handle. On more expensive kukris, the handle might be made of bone, ivory or even metal.

The scabbard is traditionally made of wood with a leather covering. There are two small pockets at the back of the scabbard to hold a pair of small knives. One of these, the *chakmak*, is for sharpening the kukri and can be used with a flint to create a spark. The other, the *karda*, is used as a penknife for skinning animals. The tip of the scabbard is protected by a metal cap. When worn in the field, the kukri is normally covered with camouflage material and attached to the soldier's webbing.

The ceremonial kukri (known as the *khonra*) is usually about twice the size of the military kukri and has a large handle to allow the user to take a double handed

When in the field, the kukri's scabbard is usually covered in camouflage material as in this photograph taken in Afghanistan in 2007

As well as being a fearsome and effective weapon, Gurkhas use their kukris as general utility tools. This photograph, taken in Afghanistan in 2007, shows a kukri being used to carve up a watermelon

A soldier from The Royal Gurkha Rifles holds his kukri aloft as he storms a trench during an infantry exercise

grip. Traditionally, the *khonra* will be used to sacrifice a water buffalo during the Hindu festival of Dashain (or 'Dashera'). The soldier selected to do this is under real pressure. If he decapitates the animal with a single blow, then his unit will be blessed with good luck in the year ahead; if he fails to do this, then bad luck will follow. Although this practice still takes place in Nepal, it is not permitted in the UK.

The kukri is ubiquitous in Britain's Brigade of Gurkhas. It has a prominent position on the capbadges of all Britain's Gurkha units and it is carried by soldiers of the brigade both on ceremonial duties and in the field. But it is not just an historic symbol. Although the basic design of the kukri has changed little over the centuries, it remains a potent weapon in the hands of a Gurkha. As Chapter 9 explains, it continues to deliver lethal effect even on today's contemporary battlefield. Its continued utility is well illustrated in this extract from the citation for the Military Cross awarded to Lance Corporal Tuljung Gurung for his actions in Afghanistan on 22 March 2013:

> Showing exceptional instinct and courage he picked up the grenade and threw it out of the sangar. The grenade detonated, peppering the

A soldier from 3rd Gurkhas about to decapitate a water buffalo in 1945 or 1946 as a sacrifice during the Hindu festival of Dashain. Severing the head with a single blow will bring the battalion good luck in the year ahead – anything less will bring bad luck

A Gurkha soldier and his kukri. In 1948, the prime minister and supreme commander of Nepal, Maharaja Padma Shamsher Jangabahadur Rana, wrote that the kukri 'is the national as well as the religious weapon of the Gurkhas. It is incumbent on a Gurkha to carry it while awake and to place it under the pillow when retiring.'[6]

A Gurkha soldier in Afghanistan with a ceremonial kukri (known as a 'khonra'). Note the two-handed grip necessary to wield this kukri

Soldiers from the Band of The Brigade of Gurkhas performing a kukri display to music

A painting by Jason Askew of the Sirmoor Battalion (later 2nd Gurkhas) defending Hindu Rao's house against an assault by mutineers during the Indian Mutiny of 1857. The kukri carried by today's soldiers is little different to that carried by the Gurkha in the centre of the painting. Then as now, it is a lethal weapon in the hands of a Gurkha

sangar with fragmentation. Gurung was again knocked off his feet. Through the obscuration of the debris he quickly identified an insurgent climbing into the sangar. Due to the close quarters, and unable to bring his rifle to bear, Gurung instinctively drew his kukri and slashed at the insurgent. In the ensuing hand to hand combat Gurung and the insurgent fell three meters from the sangar, landing on the ground outside of the Patrol Base. Exposed to possible further insurgent firing positions, he aggressively and tenaciously continued to fight with his kukri. The two insurgents, defeated, turned and fled.[7]

In addition to learning how to use a kukri as a fighting weapon, all Gurkhas also learn the martial art of Tae Kwon Do during their recruit training. Many go on to develop real expertise, gaining their black belts and competing both nationally and internationally. In November 2014, Rifleman Madhu Ghimire of 1RGR won a Gold Medal in the Army Tae Kwon Do Invitational Championship.

Other martial arts are also popular in the Brigade of Gurkhas. In October 2014, Lance Corporal Trishna Gurung from the Queen's Gurkha Signals won the Gold Medal in the under 66kg Dan Grade category in the Army Judo Championships.

A Tae Kwon Do demonstration at the Infantry Training Centre in Catterick. The Brigade of Gurkhas' view is that martial arts and the kukri make a powerful combination for close quarter fighting!

Lance Corporal Trishna Gurung of the Queen's Gurkha Signals on his way to winning a Gold Medal in the 2014 Army Judo Championships

Lance Corporal Trishna Gurung of the Queen's Gurkha Signals with the Gold Medal for winning the under 66kg Dan Grade category at the 2014 Army Judo Championships

CHAPTER 3

Fighting on the British Empire's Frontiers

1859–1913

The Gurkhas played a critical role in the Mutiny, proving their worth by remaining loyal to the British throughout the chaos that followed the uprising in Meerut. But as well as helping to restore order within India, they were also heavily involved in maintaining the security of its borders. Perhaps the most troublesome of these was that with Afghanistan, a mountainous country with the most unforgiving climate. The winters are harsh with huge snowfalls and the summers are blisteringly hot. Fighting over such difficult terrain is hard work, whether now or in the late 1830s when Gurkhas first served in Afghanistan.

Officers of the 4th Gurkhas taken in 1878 at the time of the Second Afghan War

Afghanistan first became important to the British in the early 1800s as a buffer between an expansionist Imperial Russia and an equally expansionist British Empire.[1] The main concern was that Russia would seek to invade India from Afghanistan.[2] To prevent this, the British wanted an independent but sympathetic Afghanistan as well as control of the passes that led out of the country and into British India. The problem, at least from the British perspective, was that Afghanistan was inhabited by 'some of the most warlike peoples on the face of the earth'[3] who had little intention of allowing themselves to be dominated by the British. This clash of interests led to three major wars with Afghanistan: the First Afghan War from 1839 to 1842; the Second from 1878 to 1881; and the Third in 1919.[4] Gurkhas played an important role in all three.

Gurkha involvement in the First Afghan War was limited but it is worth highlighting the background to this conflict as it set a pattern that was to repeat itself in the Second Afghan War. In the late 1830s, Lord Auckland, then governor-general of India, became convinced that the king of Kabul, Dost Mohammed, was forming an alliance with the Russians which would potentially threaten the security of British India.[5] Auckland decided that the most effective way of resolving the problem was to replace Dost Mohammed with the Amir Shah Shuja-ul-Mulk. Shah Shuja had been deposed as ruler of Kabul in 1809 and was then living in exile in British India on a pension paid by the Honourable East India Company.[6] Lord Auckland assembled an army, known as the Army of the Indus, to invade Afghanistan and restore the pro-British Shah Shuja to the throne. On 10 December 1838, the army left the then Indian city of Ferozepore and began the long trek towards Kabul.[7] Comprised of about 9500 Crown and East India Company troops, the army also included some 6000 men recruited by the East India Company but under the direct command of Shah Shuja.[8] The latter force included a battalion of Gurkha infantry as well as a contingent of Gurkha sappers and miners under the command of Major George Broadfoot.[9] The Army of the Indus encountered considerable resistance as it moved towards Kabul but eventually succeeded in routing Dost Mohammed and, on 6 August 1839, Shah Shuja entered Kabul to resume his throne in the Bala Hissar, an ancient citadel which still dominates the Kabul skyline.[10]

Major George Broadfoot who commanded a detachment of Gurkha sappers and miners during the First Afghan War

Superficially at least, the military campaign was a success.[11] The British succeeded in replacing Dost Mohammed but it was not long before widespread unrest turned into open revolt as Shah Shuja and his British supporters became increasingly unpopular. British outposts were repeatedly attacked, including the fort at Charikar which was manned by Gurkhas of Shah Shuja's army.[12] Isolated and facing certain defeat, the garrison attempted to withdraw to Kabul but, as Brigadier Christopher Bullock describes, 'it was a hopeless endeavour' and only two British officers and one Gurkha eventually made it back to Kabul.[13] The situation continued to deteriorate. The British Resident, Sir Alexander Burns, was murdered on 2 November 1841 and, by December 1841, Sir William MacNaghten, the British Envoy and Minister at the Court of Shah Shuja, believed the British position to be untenable. Accordingly, he started to negotiate a settlement with the rebel leaders, notably Akhbar Khan, the son of Dost Mohammed.[14] Unfortunately, Akhbar was not to be trusted and MacNaghten was murdered just before Christmas 1841 having gone to meet the Afghan leader to agree the settlement.[15]

A print showing the uniform of a rifleman of the 1st Gurkhas 1880, then known as the 1st Goorkha Regiment

It is conceivable that a gifted military commander might have been able to regain control of the situation. Unfortunately, the then commander-in-chief, Major General William Elphinstone, was not such a man. He had fought with Wellington at Waterloo and had assumed command in Kabul in April 1841. Although charming, he was old, ill and, according to one contemporary report, probably the most incompetent soldier of his rank in the Army.[16] Rather than try and restore order, he reopened negotiations with Akhbar and, on Christmas Day 1841, signed an agreement that was to lead to one of the most infamous retreats in British military history. It started on 6 January 1842 and lasted for just over a week. 16,500 souls began the journey from the British cantonment in Kabul;[17] only one, Surgeon William Brydon, eventually succeeded in reaching Jalalabad on 13 January 1842.[18] The British reacted swiftly to this humiliating defeat, assembling an army and retaking Kabul within the year.

The Second Afghan War started in 1878 when it was again suspected that the then king of Afghanistan, Sher Ali (one of Dost Mohammed's sons), was becoming increasingly sympathetic towards the Russians.[19] The British assembled an army, comprised of three columns and including all five of the then Gurkha regiments, and marched into Afghanistan. The bulk of the Afghan Army, which comprised about 18,000

'Storming the Peiwar Kotal' by Vereker Monteith Hamilton. The painting shows men of the 5th Gurkhas and Seaforth Highlanders attacking the left flank of the Afghan position which was at the head of a steep valley

men with 18 artillery pieces,[20] occupied a defensive position at a place called Peiwar Kotal. It was a formidable position. Located at the head of a steep mountain valley, it was 9400 ft above sea level.[21] Recognising that a frontal attack would have been destined to fail, the commander, Colonel Frederick Roberts VC, selected a small force, which included the 5th Gurkhas and the 72nd (later the 1st Seaforth) Highlanders, and carried out a night approach to try and seize the left flank of the Afghan position.[22] The attack was successful, allowing the remainder of Roberts' force to take the main enemy position.

Captain John Cook, an Edinburgh born officer who had joined the 5th Gurkhas on 27 March 1873,[23] distinguished himself throughout this operation, leading repeated charges against the enemy positions and, at one point, saving the life of a Major Galbraith, by wrestling to the ground a giant Afghan who was about to shoot Galbraith.[24] Cook was awarded the Victoria Cross for his actions. As this extract from his citation illustrates, it was well deserved:

... Captain Cook charged out of the entrenchments with such impetuosity that the enemy broke and fled, when perceiving at the close of the melee, the danger of Major Galbraith who was in personal conflict with an Afghan soldier, Captain Cook distracted this attention to himself and aiming a sword cut which the Douranee avoided, sprang upon him and, grasping his throat, grappled with him. They both fell to the ground. The Douranee, a most powerful man, still endeavouring to use his rifle, seized Captain Cook's arm in his teeth until the struggle was ended by the man being shot through the head.[25]

Captain Cook died four months later of wounds he sustained during the subsequent advance on Kabul.[26] He is buried in the British Cemetery in the Afghan capital. Perhaps fittingly, the headstone from his grave is set in a wall alongside the names of officers and soldiers from Gurkha regiments who have been killed in the British Army's more recent operations in Afghanistan.

Eventually, Roberts succeeded in occupying Kabul. Sher Ali fled the city, leaving his son, Yakub Khan, as the new but unpopular Amir. A British Resident, Sir Louis Cavagnari, was installed in July 1879 and, for a period at least, there was relative calm.[27] But it was short lived. On 3 September 1879, Cavagnari was murdered.[28] Determined to avenge

The author (left) and another officer from The Royal Gurkha Rifles laying a wreath at the headstone of Captain John Cook VC's grave in Kabul's British Cemetery in September 2014

The old palace of Shah Shuja in the Bala Hissar. Located on high ground in the centre of Kabul, the citadel dominated the city

the murder and to reassert British authority, Roberts assembled a force of some 7500 men, known as the Kabul Field Force, and marched on Kabul.[29] An Afghan army of some 14,000 tried to block him at a place called Charasiah but, repeating his tactic of using the 5th Gurkhas and the Highlanders to conduct a surprise flanking attack,[30] Roberts out manoeuvred the Afghans and, by mid October 1879, he was once again back in Kabul. Interestingly, when Roberts was eventually created a peer, he included a highlander from the 72nd and a rifleman from the 5th Gurkhas on his coat of arms to commemorate the lasting relationship he had formed with these two regiments.[31,32]

Once in Kabul, Roberts began consolidating his position. Martial law was imposed, heavy fines were levied and those responsible for Cavagnari's death were executed.[33] Perhaps understandably, the Afghans reacted by calling for a Jihad, or Holy War, against the invaders and, at dawn on 23 December 1879, Roberts and his 7000 men found themselves being attacked by a force of some 50,000 Afghans.[34] Despite their numbers, the Afghans were unable to penetrate Roberts' defences,

Officers and men of the 4th Gurkhas posing for a picture in the Bala Hissar

largely because they were hopelessly outgunned; not only did Roberts' force have modern breech loading firearms but they also had two Gatling machine guns and 22 artillery pieces.[35] The Afghans eventually dispersed having sustained 3000 casualties; by contrast, the defenders apparently sustained only 33.[36]

In Spring 1880, Roberts was joined in Kabul by Lieutenant General Sir Donald Stewart and his force of some 7200 men. Stewart's journey up from Kandahar had not been without incident. At one point, his advance guard had been attacked by a force of some 15,000 Afghans. In the ensuing battle, known as the Battle of Ahmed Khel, the 3rd Gurkhas had distinguished themselves by repeatedly halting the advances of the Afghan cavalry.[37] But the Afghans also achieved some successes. On 27 July 1880 at the Battle of Maiwand, they destroyed a brigade sized force and then laid siege to the survivors in the fortress of Kandahar.[38] Roberts responded as quickly as he could by assembling a force of some 10,000 men and marching the 280 miles from Kabul to Kandahar in 20 days,[39] a quite remarkable achievement given the terrain. When he arrived in Kandahar, Roberts found that Ayub Khan, a claimant to the throne, had assembled an army of some 25,000 and had occupied a defensive position on the high ground dominating the city.[40] The 4th Gurkhas and 2nd Gurkhas, fighting alongside the Gordon Highlanders, fought with distinction in the battle which ensued. During the battle, a young rifleman from the 2nd Gurkhas reputedly distinguished himself by ramming his hat down the barrel of an Afghan artillery piece, claiming it for his regiment![41]

A painting of the Battle of Kandahar showing the capture of an Afghan gun by a young Rifleman of the 2nd Gurkhas

The Battle of Kandahar was the final battle of the Second Afghan War but skirmishes along Afghanistan's border with British India would continue well into the next century. Perhaps inevitably given their obvious talent for mountain warfare, the Gurkhas would remain intimately involved – indeed, for the 5th Gurkhas, who became members of the Punjab Frontier Force, the North West Frontier was an almost permanent posting.[42]

Some of the routine campaigns that were conducted along the border were significant. The Black Mountain Campaign of 1888, for example, involved the deployment of some 9500 men as part of the hastily assembled Hazara Field Force.[43] The force deployed on 4 October 1888 to punish rebellious tribesmen for killing two British officers and four Gurkhas whilst they were conducting a reconnaissance patrol in the Black Mountains, a remote border area at the northern end of the Punjab.[44] Over the next month or so, the force killed between 150 and 200 tribesmen as it made its way deeper into the inhospitable mountains. After the destruction of several villages, the tribesmen agreed to British terms and, on 13 November 1888, the Hazara Field Force disbanded.[45] British losses totalled 32 dead (seven from disease) and 54 wounded.[46]

Described as 'something of a forgotten colossus', the Tirah Campaign of 1895 is also worth highlighting.[47] The British assembled an army of 44,000 men in order to regain control of the Khyber Pass and to retake

Gurkhas from the Hazara Field Force forcing their way deeper into the Black Mountains in 1888. The near vertical nature of the terrain made life difficult, even for the pack-mules used to carry supplies

Soldiers from 2/5th Gurkhas returning fire during an engagement in the Black Mountain Campaign of 1888

areas of British India that had been seized by Pathan tribesman.[48,49] The main difficulty from the British perspective was that the Afghans were wily operators and generally refused to meet head on, preferring instead to snipe from a distance and then melt away before they could be decisively engaged. Interestingly, this same tactic would be employed by the Taliban against both Coalition and Afghan National Security Forces (ANSF) over 100 years later.

The Tirah Campaign lasted for almost 18 months. Arguably the most significant engagement of the campaign occurred on the heights surrounding the village of Dargai.[50] Heavily defended, the village could only be reached by an exposed mountain track which ended in open ground immediately below the village.[51] On 19 October 1897, the 3rd

Gurkhas assaulting the enemy position at Dargai. The painting shows Piper George Frederick Findlater of the Gordon Highlanders who, despite being shot in the feet and exposed to enemy fire, continued to play, encouraging the assaulting force. He was awarded the Victoria Cross for his heroic actions

Gurkhas and King's Own Scottish Borderers succeeded in taking the village but, for reasons which are still not entirely clear, they decided not to remain in position overnight. When the 2nd Gurkhas advanced to reoccupy the village on the morning of 20 October 1897, they found to their cost that overnight hundreds of well-armed tribesmen had infiltrated back into the village.[52] It took determined charges by the 2nd Gurkhas and the Gordon Highlanders, joined by the Dorsets, the Derbyshires and the 3rd Sikhs, to eventually retake the village.[53] The casualties were significant: over the three days of the Dargai operation, the 2nd Gurkhas lost 17 killed and 48 wounded; the 3rd Gurkhas lost 17 killed and 46 wounded.[54]

We will return to Afghanistan in Chapter 5 and again in Chapter 9 but, at the same time as protecting its North West Frontier, British India also had to protect its northern and eastern borders. Again, Gurkhas played a significant role in the many campaigns and skirmishes that characterised life on the fringes of the British Empire. Acts of remarkable daring and bravery were commonplace. Three particular events, each of which led to protagonists being awarded the Victoria Cross for their actions, provide an insight into the sort of military activity that occupied Gurkha battalions in the latter half of the 19th century.

The Indian state of Assam lies on the southern part of the then border with Burma. The terrain is largely comprised of mountainous jungle but, when combined with the climate, it is ideal for growing tea. Perhaps inevitably, as British tea planters sought to expand their plantations, they found themselves rubbing up against the indigenous tribes whose land they were trying to cultivate. The tribesmen, particularly the Lushai and the Nagas,[55] were keen to defend their land and murderous raids on the tea plantations were common.[56] In one particular raid on 23 January 1871, a young girl called Mary Winchester was abducted by Lushai tribesman.[57,58] It, more than any other outrage committed by the Lushai, appeared to catch the public imagination. Efforts to negotiate Mary's release, as well as that of the many other hostages taken by the Lushai, failed.[59] Eventually, a large expedition was dispatched in late 1871. Its military commanders were '... specifically instructed that the object of the expedition was not one of pure retaliation, but that the surrender of the British subjects held in captivity should be insisted on.'[60] In his memoir of the then Major General Sir Frederick Roberts VC, the historian Charles Rathbone Low notes that 'to recover this second little Helen of Troy ... a considerable force was assembled, and the Indian Government found itself involved in hostilities.'[61]

2/2nd Gurkhas on parade at the start of the Lushai expedition of 1871/1872. Major Donald Macintyre of the 2nd Gurkhas was awarded the Victoria Cross during the campaign for his actions at the village of Lal Gnoora and Rifleman Inderjit Thapa was awarded the Indian Order of Merit (IOM), Third Class

Battalions from both the 2nd and 4th Gurkhas took part in this enterprise, which was to become known as the First Lushai Expedition.[62] By 3 January 1872, the 2nd Goorkhas had located the Lushai village of Lal Gnoora.[63] The village was well defended and surrounded by nine-foot-high bamboo stakes, known as panjis, which, sharpened at one end and stuck in the ground, made fearsome obstacles.[64] The attacking force was engaged as it approached the village, taking heavy casualties.[65] The Lushai had deliberately set fire to houses in the village and the smoke added to the confusion.[66] But it also created an opportunity and, seizing this, Major Donald Macintyre, then the second-in-command of the 2nd Gurkhas, ran forward using the smoke as cover and, with Rifleman Inderjit Thapa, succeeded in clambering over the bamboo pickets.[67,68] Their sudden appearance in the midst of the burning village startled the Lushai and, before long, the position had been taken. Major Macintyre was awarded the Victoria Cross for his bravery with Rifleman Inderjit Thapa receiving the Indian Order of Merit (IOM), Third Class.[69] Many British hostages were released during the Expedition, including Mary Winchester.[70]

In March 1891, the pro-British ruler of the independent state of Manipur, located in the south east of Assam, was deposed by his brothers.[71] The

Because of the steep terrain, elephants were often used to pull artillery pieces during campaigns on the borders of British India. This sketch shows the 2nd Gurkhas helping man-handle guns on the Lushai Expedition

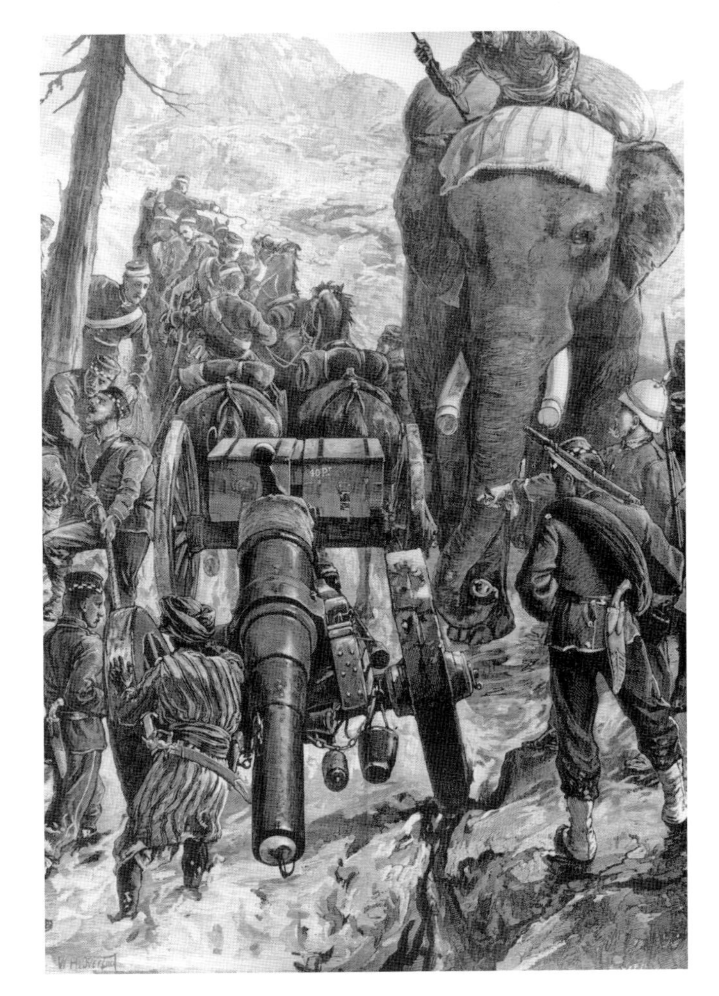

Gurkha officers of the 8th Gurkhas in 1886, then known as the 44th Regiment Goorkha (Light) Infantry

British commissioner in Assam, James Wallace Quinton, dutifully set off to arrest the ringleaders.[72] He was escorted by Lieutenant Colonel Charles Skene, the commanding officer of the 42nd (later 6th Gurkhas), and 454 men drawn from both the 42nd and 44th (later 1st Battalion 8th Gurkhas).[73,74] The delegation arrived in Imphal, the capital of Manipur, but violence broke out when Colonel Skene tried to arrest one of the ringleaders.[75] After two days of fighting, Commissioner Quinton and Lieutenant Colonel Skene, along with Frank St Clair Grimwood (the British political agent in Manipur), left the relative safety of the British Residency to discuss a truce with the Manipuris.[76] It was a fatal mistake. The Manipuris, who had no intention of making peace, quickly seized and murdered the delegation.[77] According to one contemporary account, their bodies were dismembered and their limbs '... were then thrown outside the city walls to be devoured by pariah dogs.'[78]

There are several accounts of what then happened to those who remained trapped in the Residency. One account is that they managed to fight their way out of the capital, eventually meeting up with a column of 200 Gurkhas from the 43rd (later the 2nd Battalion of the 8th Gurkhas) on its way to support them.[79] Another is that the two surviving British officers, Captain Louis Boileau and Captain George Butcher, fled the residency with Mrs Ethel Grimwood, the very beautiful wife of the murdered political agent, leaving the Gurkhas still surrounded in the Residency to fend for themselves.[80] Whatever the truth, in November 1891, Captains Boileau and Butcher were '... removed from the Army, in consequence of their conduct during the recent disaster at Manipur, when a number of British officials were massacred by the natives.'[81]

Lieutenant Charles Grant, commanding a detachment of the 43rd (later 2nd Battalion, 8th Gurkhas) 50 miles away in Tamu, was made of much sterner stuff.[82] Immediately he heard of the troubles in Manipur, he sought permission to go and help his beleaguered countrymen. On 28 March 1891, at the head of his small detachment of 80 men from the 43rd and supported by three pack elephants carrying supplies, he set off for Manipur.[83,84] But his route to the capital was blocked by about 800 Manipuris at a place called Lang Thobal. Grant didn't falter. He ordered his Gurkhas to fix bayonets and charged at the Manipuris who, startled by this unexpected and remarkably courageous behaviour, withdrew.[85,86] The Manipuris counter-attacked in greater force the next day. Undaunted, Grant defended his position, only withdrawing ten days later when ordered to do so by his superiors.[87] Grant was awarded the Victoria Cross for his actions and every surviving member of his

The surviving Gurkhas from Lieutenant Charles Grant's detachment who were all awarded the Indian Order of Merit for their bravery against a force of some 800 Manipuri tribesman at Lang Thobal. Lieutenant Grant was awarded the Victoria Cross for his actions

brave detachment received the Indian Order of Merit.[88] Returning to the North East Frontier, the atrocities committed by the Manipuris outraged British India.[89] An army of 5000 was quickly assembled to bring the Manipuris to heel and, within a few weeks, peace was restored.

In December 1903, Colonel Francis Younghusband, then aged 40 and a veteran of campaigns on the North West Frontier,[90] led a military expedition into Tibet to try and consolidate relations between British India and the Dalai Lama. The main driver for the expedition was a belief that Russia was gaining too much influence with both China and Tibet and that this might lead to the latter's annexation by Russia,[91] threatening the security of British India. Although intended to be peaceful in nature,[92] Younghusband's expedition was aimed at forcing the Tibetans to the negotiating table. The expedition comprised of some 3000 soldiers and included a detachment of six companies from the 1st Battalion of the 8th Gurkhas (then known as the 8th Gurkha Rifles).[93]

After a number of minor skirmishes, the expedition found its route to Lhasa, the Tibetan capital, blocked by a formidable fortress at Gyantse.[94] The fortress, which was manned by some 6000 Tibetan troops, was built on a sheer rock outcrop which rose 600 ft from the valley below.[95] On 6 July 1904, the British started their assault but the fortress seemed impregnable.

Colonel Francis Younghusband's bodyguard, which included men from the 8th Gurkhas, during the expedition into Tibet of 1903

The route taken by Lieutenant John Duncan Grant and Havildar Karbir Pun in scaling the cliffs to get into the Tibetan fortress at Gyantse. Lieutenant Grant was awarded the Victoria Cross and Havildar Karbir Pun received the Indian Order of Merit, First Class for their actions

The main body assaulting the formidable Tibetan fortress at Gyantse in 1904

Eventually, Lieutenant John Duncan Grant and Havildar Karbir Pun of the 8th Gurkhas managed to scale the sheer walls of the cliff, despite being subjected to constant enemy fire, and force their way into the fortress. It was a turning point in the attack and the position was swiftly taken by the remainder of the force. For their courage in the face of the enemy, Lieutenant Grant was awarded the Victoria Cross and Havildar Karbir Pun received the Indian Order of Merit, First Class.[96] The expedition eventually reached Lhasa on 3 August 1904.[97] Although the Dalai Lama had left the city, an agreement between Tibet and the British, known as the Lhasa Convention, was eventually signed on 7 September 1904.[98] Interestingly, Lieutenant Grant's Victoria Cross was auctioned on 2 July 2014 for a record £408,000. The medal was bought by Lord Ashcroft and is now on display at the Imperial War Museum.

Hill men from birth, the Gurkhas excelled at fighting against unruly but wily tribesmen in the mountains on India's borders. Their physical robustness and courage enabled them to make an invaluable contribution to the Empire's security. But they were soon to find themselves fighting an entirely different sort of war as the storm clouds gathered over Europe in the build up to the Great War.

Mounted Gurkha infantry raised circa 1898 with the role of dashing ahead to clear the advance route for the main infantry force. The concept did not prove to be a great success and the force was disbanded in 1902

Hill Racing in the Brigade of Gurkhas

Over the years, Gurkhas have established a reputation as remarkable hill runners. Their powerful legs and low centre of gravity give them an edge over taller runners, particularly downhill. Hill Racing has a long tradition in the Brigade of Gurkhas and was first introduced by Major The Honourable (later Major General) Charles Bruce of the 5th Gurkhas to demonstrate that Gurkhas could compete favourably with Punjabis, Sikhs and the other Indian castes recruited into the Indian Army.[1]

The Khud Race teams of the 5th Gurkhas and the 3rd Gurkhas circa 1908

Runners from the 5th Gurkhas and the 3rd Gurkhas compete in the annual Khud Race in the 1930s. As this picture shows, the route is usually extremely steep and requires strong nerves as well as powerful legs!

In 1890, Hill Racing, or 'Khud Racing' as the Gurkhas call it, was formally introduced as an event in the Punjab Frontier Sports Competition.[2] There were 133 entrants. To the surprise of many present, the first 33 places were won by soldiers from the 1st and 2nd Battalions of the 5th Gurkhas, with the first Punjabi coming in 34th.

In 1894, the annual Hill Race was opened to all comers from across the Brigade of Gurkhas. Initially, only the 5th Gurkhas and 3rd Gurkhas took part but, by 1901, the race was becoming increasingly well-established and the 6th Gurkhas were starting to dominate. Traditionally, the winning regiment selected where next year's event would be held. Arguably this gave them an advantage and might help explain why the best runners appeared able to maintain their dominance for a number of years. Rifleman Budhiparsad Thapa of 1/3rd Gurkhas, for example, won the race in 1895, 1896, 1897, 1898, 1900 and 1901. Rifleman Dharmjit Pun of 1/6th Gurkhas won in 1902, 1903, 1904 and 1906. Tulbir Pun of 1/6 th Gurkhas won the race 5 times, initially as a Rifleman, then as a Lance Naik and lastly as a Naik in 1909, 1910, 1911, 1912 and 1913.

A typical course would take a top runner between 23 to 33 minutes and would include a climb of well over 1000 ft. Occasionally, sceptical individuals would doubt the Gurkhas' ability as hill runners with interesting results. One such incident occurred on the Isle of Skye in 1899 when Major Charles Bruce and Havildar Harkabir Thapa, both experienced mountaineers,

The 6th Gurkhas Khud Race team taken sometime between 1909 and 1913. The legendary Tulbir Pun, who won the race five times over this period, is shown seated on the ground in the centre holding the Hill Race Trophy (known as 'Little Man')

The start of the Khud Race up 'Nameless Hill' in Hong Kong's New Territories. This photograph was taken in 1986. Ten years later, the last Khud Race was held in Hong Kong prior to the former Colony being handed over to the Chinese Government

were travelling in the Chuchullin Hills. The local laird, McLeod of McLeod, bet Major Bruce that Harkabir could not run from the Sligachan Inn to the top of Mount Glamaig and back in an a hour and a quarter. This seemed a reasonable bet given there were two miles of open moorland to the foot of the mountain and then a climb of 2817 ft to the summit. The laird's ghillies, egging on their master, believed that even their dogs couldn't do this. To the surprise of his doubters and the laird in particular, Harkabir won the bet, taking only 37 minutes to reach the summit and only 18 to get back to the inn. This record survived until 1997 when it was eventually broken by a professional fell runner! Harkabir continued to serve in the Gurkhas. Although wounded at Gallipoli, he survived the Great War and retired as a subedar major.

Over the years, the Hill Race has taken various forms from individual best effort through to a relay involving teams of eight men. Races were suspended for several years during both world wars and during the Malayan Emergency. When the Brigade of Gurkhas was based in Hong Kong, the race was an important annual event and proved to be hugely popular with both competitors and spectators. It was keenly fought with Gurkha teams following intensive training programmes for months beforehand to try and give them an edge over their competitors. The last formal Hill Race in Hong Kong took place in 1996 and was organised by 48 Gurkha Infantry Brigade. After the Brigade of Gurkhas relocated to the UK, the Hill Race declined in popularity. However, in May 2014, the Second Battalion of The Royal Gurkha Rifles held the first Hill Race for nearly 20 years.

A runner from 3rd Gurkhas on his way to the finish line. The picture was taken in the 1930s

CHAPTER 4

The First World War

1914–1918

The First World War was a defining event for the Gurkhas, not just because it was the first time that these proud warriors had fought on European soil[1,2] but also because it was their first exposure to 'modern' warfare. Despite being poorly prepared and ill equipped for the horrors of the trenches, they were to make a lasting impression, winning three Victoria Crosses and numerous other gallantry awards. But the reputation they earned and the many accolades they received came at a cost; of the 90,780 Gurkhas who fought for the Crown during the Great War,[3] 20,000 became casualties. Of these, 6168 died. This chapter highlights some of the actions in which the Gurkhas played a key role, both on the Western Front and further afield on the Gallipoli Peninsula and in Mesopotamia.

There were ten Gurkha Regiments at the outbreak of the war, each comprised of two battalions.[4] As we saw in the last chapter, up until this point

Gurkhas sharpening their kukris in France. Although the picture was taken about a hundred years ago, the kukri that Gurkhas use on contemporary operations remains virtually unchanged from those shown in the photograph

A Gurkha in full winter trench kit. Although the photograph was taken in 1914, many units of the Indian Corps deployed to the Western Front at the outbreak of the war without winter clothing and equipment

they had been primarily involved in maintaining the security of British India. They were therefore equipped and trained for frontier operations against small groups of determined tribesmen rather than for massed attacks against highly disciplined, well equipped European troops. Two divisions deployed from India to support British operations in Europe, both containing Gurkha units.[5] The Meerut Division had four Gurkha infantry units (1/9th, 2/2nd, 2/3rd and 2/8th Gurkhas) whilst the Lahore Division contained only two (1/1st and 1/4th Gurkhas).[6] Both divisions formed part of the Indian Corps which was deployed to reinforce the British Expeditionary Force (BEF) to the south of Ypres.[7]

On 29 October 1914, 2/8th became the first Gurkha unit to be sent into the line.[8] It was a sobering experience; within 24 hours, the Germans had attacked, inflicting heavy casualties. By the evening of 30 October 1914, four British officers, four Gurkha officers and 146 Gurkha other ranks from the battalion had been killed; a further three British officers and 61 Gurkha other ranks had been wounded.[9] 2/2nd found themselves in similar circumstances several days later. Along with 1/9th Gurkhas, they were deployed into defensive positions on the forward edge of the village of Neuve Chapelle. After a fierce artillery bombardment, the Germans attacked, seizing some of the trench positions on 2/2nd's flank. 1/9th tried to retake the position but the weight of German fire was too great. 2/2nd then tried to counter-attack but the Germans were too strong. Again, the Gurkha casualties were high with 2/2nd losing seven British officers, four Gurkha officers and 33 men killed; a further 99 Gurkha other ranks were either wounded or missing.[10]

2nd Gurkhas with kukris drawn training to attack the German trench system at Neuve Chapelle

The Germans retook Neuve Chapelle and the Gurkha battalions, like the other infantry units on the front line, settled into the routine of manning the trenches or being taken 'out of the line' to rest. The routine was occasionally broken as one or other side tried to seize the initiative. Though infrequent, these actions could be costly. 2/8th, for example, lost one Gurkha officer and 22 men killed, with a further five British officers, one Gurkha Officer and 22 men wounded, in heavy fighting near the village of Festubert.[11] Notably, Subedar Shamsher Gurung, who was severely wounded during this exchange, received the Order of British India for his courage in crawling back to friendly positions over a period of three days.[12] On 13 November 1914, 2/3rd Gurkhas sustained heavy casualties in a failed night raid near the village of Richebourg.[13] The men of 1/9th Gurkhas, who were supporting them, distinguished themselves by going out into no-man's-land to help recover their wounded countrymen back to safety.[14]

But the stalemate could not endure and, on 10 March 1915, the Allies launched a major offensive to try and retake Neuve Chappelle. The plan was to seize the village and then exploit beyond it to secure the Aubers

2/3rd Gurkha Rifles in France. Like other infantry battalions, the Gurkhas alternated between service in the trenches and time 'out of the line' to recuperate

Gurkhas from 1/9th Gurkhas in the trenches of Flanders in 1915. The Gurkha on the right is using a periscope to observe the area to his front without the risk of being shot

Ridge which, heavily defended by the Germans, dominated the Allied positions.[15] 2/3rd Gurkhas, as part of the Garhwal Brigade, were in the vanguard of the attack. As soon as the Allied artillery barrage lifted, they raced over the open ground separating them from the German trenches and, kukris drawn, secured the position, capturing Neuve Chapelle.[16] It was the first time on the Western Front that the German line had been broken and was,[17] as Brigadier Christopher Bullock notes, a 'brilliant success.'[18] The follow on force, which comprised the Dehra Dun Brigade and included 2/2nd and 1/9th Gurkhas, eventually pushed forward towards the Aubers Ridge but, as darkness fell, they were forced to occupy a hasty defensive position short of the ridgeline. The Germans then launched a massive counter-attack to try and retake Neuve Chapelle.[19] Comprised of some 16,000 men, it was a formidable force but the Dehra Dun Brigade held firm, inflicting some 3000 casualties on the enemy force.[20] There were numerous acts of heroism amongst the Gurkhas at Neuve Chapelle. Rifleman Gane Gurung, for example, singlehandedly captured eight German soldiers whilst clearing one of the houses in the village.[21] He was awarded the Indian Order of Merit for his bravery.[22]

Rifleman Kulbir Thapa of 2/3rd Gurkhas who earned a Victoria Cross for his actions at the Battle of Loos in September 1915, becoming the first Gurkha to receive the award.

On 25 September 1915, the Allied forces advanced in what would become known as the Battle of Loos. It was the 'big push' expected to break the German line. As ever, the Gurkhas were in the thick of it with both 2/8th and 2/3rd forming part of the assaulting Garhwal Brigade.[23] Both battalions sustained dreadful casualties. 2/8th advanced and managed to clear a number of German trenches but their victory was short lived as, isolated from the units on their flanks, they had little option but to fall back.[24] In the single day's fighting, 2/8th's casualties amounted to nine British officers, eight Gurkha officers and 453 men; a further 166 men, many of whom later died, were taken prisoner.[25] 2/3rd were mown down by German machine guns as they tried to cross wire obstacles which the British artillery barrage had failed to breach. An officer and 38 men were sent out to cut a way through the wire but were all killed, less for Rifleman Kulbir Thapa. Although badly wounded, he started to crawl back towards the British positions but came across a soldier from the 2nd Leicestershire Regiment who was also severely wounded.[26] The British soldier urged Kulbir to save himself but Kulbir remained with the wounded man throughout the night and, early the next morning, started to carry him towards the Allied lines.[27] He then stumbled across two wounded Gurkhas, both unable to walk.[28] He dragged the British soldier into cover and then went back for the

Gurkhas. One at a time, he carried them to the safety of the Allied lines before returning to collect the British soldier. All three survived and Kulbir became the first Gurkha to be awarded the Victoria Cross for his conspicuous bravery.

The Indian Corps was eventually withdrawn from operations in Europe in the late Autumn of 1915. By then, additional British and Canadian troops had begun to arrive on the Western Front and it was decided that the Indian Corps' infantry would be better employed in Egypt, East Africa and Mesopotamia where the '... climate and general conditions would be more familiar to them, and contacts with India much easier.'[29] The Corps' arrival on the Western Front in late October 1914 had prevented the Germans from breaking through the British defences and reaching the Channel ports.[30] Its actions at Neuve

1/1st in France in 1915. Despite the horror of life in the trenches, the smiles in this photograph illustrate the bond that existed, then as now, between British and Gurkha Officers

Chapelle had also clearly demonstrated that the seemingly invincible Germans could be driven from their trenches.[31] By the time it left the Western Front, the Indian Corps had sustained 25,000 casualties and nine of its members had been awarded the Victoria Cross. Although many regiments within the Corps distinguished themselves, the Gurkhas in particular established a reputation as the most fearsome and loyal of soldiers – indeed, General Sir James Wilcox, then commander of the Indian Corps, was unequivocal that the Gurkhas were his best soldiers.[32,33]

In April 1915, a force of some 75,000 soldiers from Britain, France, New Zealand and Australia landed on the Gallipoli Peninsula in the south of Turkey.[34] The intention was to seize the ground dominating the Dardanelles, the strip of water that connects the Mediterranean to the Black Sea, in order to open a sea route to Russia and to bring Constantinople (now Istanbul) within range of British warships.[35] A previous attempt to take the straights with a purely naval force had failed after three British ships had hit mines and sunk.[36] General Sir Ian Hamilton, the commander of the Mediterranean Expeditionary Force, had asked for 100,000 men for the land operation but had been given far fewer. An accomplished soldier, he was a veteran of the North West Frontier and recognised the value that Gurkhas would add in the hilly terrain of the Peninsula. He wrote to Lord Kitchener, then secretary of state for war, in March 1915 requesting that he be given a brigade of Gurkhas.[37] He got his wish and 29 Indian Infantry Brigade, which included three battalions of Gurkhas (1/5th, 1/6th and 2/10th), joined his force, albeit several days after the initial landings had taken place. 1/6th Gurkhas were the first battalion to arrive and had immediate impact, seizing a position which two British units had been unable to take.[38] So impressed was Hamilton by the battalion's performance that he renamed the feature Gurkha Bluff.[39] Hamilton then switched his focus to a feature known as Achi Baba which, lying to the north of Gurkha Bluff, dominated the beaches on which the Allied forces had landed.[40] All three of 29 Indian Infantry Brigade's Gurkha battalions were committed to trying to take the feature but the Turks, who also appreciated its tactical importance, held firm, inflicting tremendous casualties on the attacking forces.[41] 1/5th lost 129 men and seven British officers within the first few hours of the attack with 1/6th sustaining 95 casualties.[42]

In a separate operation, 2/10th achieved a notable victory at Gully Ravine, scaling a sheer cliff to surprise the defending Turks.[43] But notwithstanding the occasional success, the constant fighting took its

Soldiers from the 6th Gurkhas in the trenches of Gallipoli in 1915

toll on the Gurkhas who were always in the thick of it. Within 35 days of arriving at Gallipoli, for example, 2/10th had lost three quarters of its British officers and 40 per cent of its other ranks.[44] The Brigade was therefore pulled out of the line and given a month to recuperate on the Isle of Imbros,[45] an Allied staging post for operations on the Gallipoli Peninsula.

By August 1915, 29 Indian Infantry Brigade was back in action, this time as part of a new offensive further to the north in the ANZAC (Australian and New Zealand Army Corps) area of operations. The plan was to seize the central Sari Bair massif in order to gain control of the Peninsula and isolate the Turkish forces which were causing so many problems down in the south. As Hamilton notes 'the first step in the real push – the step which above all others was meant to count – was the night attack on the summits of the Sari Bair ridge.'[46] But the night attack did not go as planned and, as dawn broke on 9 August 1915, Allied troops had still not reached the crest of the ridgeline. 1/6th Gurkhas, under the command of Major Cecil Allanson, were eventually launched at the Sari Bair feature, the highest of the ridgeline's peaks.[47] The fighting was intense, with Gurkhas drawing their kukris and using their weapons as clubs as the ammunition ran out. Eventually, supported by two companies of the South Lancashires, 1/6th reached the top of the peak, driving the Turks down the far side.[48] The Gurkhas pursued

Major Cecil Allanson under whose inspirational leadership the 6th Gurkhas succeeded in capturing the critical point of the Sari Bair massif on 9 August 1915. A remarkable officer and superb athlete, Major Allanson held the Army record for the 2 miles for a number of years. He was recommended for the Victoria Cross for his actions at Sari Bair but received the Distinguished Service Order (DSO)

them until, being mistaken for fleeing Turks, they were engaged by the guns of HMS Colne.[49]

Once part of the ridgeline had been secured, the plan was that four battalions, under the command of General A H Baldwin, would then use the lodgement to exploit along the ridge, clearing the enemy's positions. But Baldwin's battalions had lost their way during the night approach and never arrived.[50] The Turks quickly counter-attacked, pushing the Gurkhas and reinforcements from the South Lancashires and the Wiltshires, off the ridgeline and pinning them down on the mountainside.[51] By this stage of the battle, all of 1/6th's British officers, less the medical officer (Captain Edward Phipson), were either dead or wounded and it

A painting by Terence Cuneo showing 1/6th Gurkhas assaulting well prepared Turkish positions on the highest feature of the Sari Bair massif. The painting shows Major Cecil Allanson leading his men from the front

was left to Subedar Major Gambirsing Pun to command the battalion as it withdrew. The next day, the Turks counter-attacked in force, consolidating their defensive positions on top of the ridgeline and driving the Allies back down towards the beaches. Though wounded, Major Allanson survived the attack on Sari Bair. He was recommended for the Victoria Cross but eventually received the Distinguished Service Order (DSO) for his actions.[52]

The outcome was perhaps not entirely unpredictable. Several days before the operation, Allanson had written in his diary that '... the more the plan was detailed as the time got nearer, the less I liked it, especially as in my own regiment there were four officers out of seven who had never done a night march in their lives.'[53] He very candidly asked 'what would one have done to a subaltern at a promotion examination who made any such proposition?'[54] The Allied forces sustained 12,000 casualties in the failed operation but the number of casualties was to increase. The Allied trenches were within Turkish artillery range and this, combined with the harsh winter conditions, ensured that many more soldiers died in the months that followed. By October 1915,[55] it was apparent that additional forces would be necessary to defeat the Turkish defenders. As these could not be spared, Kitchener made the decision to withdraw from the Peninsula.[56] In mid January 1916, and now under the command of General Sir Charles Monro, the last of the Mediterranean Expeditionary Force left Gallipoli.[57] Over the course of the campaign, the Allied force had sustained some 205,000 casualties. The French had fared particularly badly, sustaining 47,000 casualties out of 79,000 men.[58]

Gurkhas also found themselves fighting the Turks in Mesopotamia (modern day Iraq) as part of a large British force deployed to secure access to Persia's oilfields.[59] Advancing up from the Persian Gulf, the British seized Basra with little difficulty in November 1914 and then, on 14 April 1915, defeated a large Turkish force of some 12,000 at Shaiba to the south-west of Basra.[60,61] The British force then divided into two. Major General George F Gorringe, commanding the 12th Division, advanced following the line of the Euphrates whilst Major General Charlie Townshend, commanding the 6th Indian (Poona) Division, followed the Tigris.[62] Townshend made swift progress, occupying Amara on 3 June 1915 after using a flotilla of boats to carry his troops up the Tigris.[63] Gorringe's route proved more difficult. A large proportion of his force, which included 2/7th Gurkhas, succumbed to dysentery, malaria and other waterborne diseases as they made their way up the Euphrates towards the Turkish defences at Nasiriya.[64] Not only was the position well defended but much of the area was flooded, adding to the complexity of Gorringe's task. Eventually, on 24 July 1915 and after a month-long campaign, the town fell to the British.[65] Although down to about a half of their strength, 2/7th Gurkhas played a key role in the defeat by breaking into the Turkish trench system and, using bayonet and kukri, routing the defenders.[66]

The British had made surprisingly good progress and it was therefore decided to continue advancing north towards Kut-al-Amara,[67] a major Turkish stronghold about 200 miles east of Baghdad.[68] Townshend's daring plan enabled him to out-manoeuvre the Turkish defenders and, on 28 September 1915, he succeeded in taking the city, capturing 1300 prisoners and driving the defeated Turks northwards out of Kut.[69,70] Following these successes, General Nixon believed that his force could continue on and secure Baghdad. General Townshend had severe reservations, believing that at least two divisions would be needed to take the capital.[71] After much discussion, on 23 October 1915 Nixon received permission to march on Baghdad and, on 14 November, the advance began.[72] Unbeknownst to the British, however, the Turks had received significant reinforcements and some 18,000–21,000 Turks and Arabs, supported by 45 artillery pieces,[73] occupied a well prepared defensive position amongst the ruins of the ancient city of Ctesiphon, 22 miles south of Baghdad.[74,75] Townshend's force, which comprised about 13,000 British and Indian troops, launched their attack on 22 November 1915.[76,77] They broke through the first line of trenches but were unable to breach the second. At one point, some 300 men of 2/7th Gurkhas, along with about 100 men from the 24th Punjab Regiment, found themselves isolated on a raised mound and surrounded by a Turkish division.[78,79] On 26 November 1915, outnumbered, outgunned and having sustained losses of some 4600 men, Townshend ordered his force to withdraw back to Kut-al-Amara.[80]

Soldiers from 6th Gurkhas in Mesopotamia in 1917. Note the use of the periscope to observe the area to the front without being visible to the enemy

The Turks followed in hot pursuit. Surrounded on three sides by the Tigris, Kut was in many ways an ideal place for Townshend's force to await the arrival of a relief column. However, despite repeated efforts, which involved 1/1st and 1/9th Gurkhas, the British were unable to break through the Turkish cordon. Attempts were made to resupply Townshend's beleaguered division by aircraft, the first time this had ever been done,[81] but it was a token effort given the size of Townshend's force. In April 1916 and using T E Lawrence (Lawrence of Arabia) as an intermediary, the British even offered the Turkish commander £2 million to let Townshend's force go free but the offer was 'disdainfully refused.'[82] On 29 April 1916, Townshend surrendered unconditionally to the Turkish General Khalil Pasha.[83] As John Parker notes 'and so it was that 2600 British and 10,486 Indian Gurkha troops were volunteered into the custody of the Turks.'[84]

In August 1916, General Frederick Maude was appointed as the commander of the Army in Mesopotamia; on 13 December 1916, he went on the offensive, resuming the advance to Baghdad at the head of an army of some 165,000 men.[85] About 110,000 of his force were Indian and Gurkha and included 1/2nd, 4/4th, 1/7th, 2/9th and 1/10th.[86] Following Townshend's surrender, the Turks had occupied Kut, preparing defensive positions either side of the Tigris. Maude realised that

Soldiers from 9th Gurkhas prepare for the crossing of the Tigris in February 1917 in order to defeat the Turkish defences at Kut. The operation was made particularly difficult because not only was the river in spate following heavy rains but the opposing banks were covered by enemy machine guns

he would need to clear Kut before he could continue the advance to Baghdad. His plan was to advance up both sides of the river and then, having pushed the Turks back on the west bank, move additional forces across the river in order to attack the rear of the Turks' main defensive position.[87] It was an audacious plan, particularly as the river had swollen following heavy rains. On the morning of 23 February 1917, D Company of 2/9th Gurkhas, commanded by Major George Campbell Wheeler, and a detachment from 1/2nd, commanded by Lieutenant C G Toogood, succeeded in establishing tentative footholds on the opposing bank despite heavy enemy resistance.[88] The Norfolk Regiment also managed to secure a foothold in a slightly less exposed position, allowing the remainder of 2/9th and 1/2nd to cross the river and expand the bridgehead. By early afternoon, a boat bridge had been established, allowing Maude's forces to attack the rear of the Turkish defences. Realising that their position had become untenable and recognising that they were in danger of being cut off,[89] the Turks withdrew towards Baghdad. It had been a brilliant victory but it had come at a cost with 1/2nd and 2/9th losing 98 killed and 132 wounded.[90] For his conspicuous bravery, Major Wheeler, who led with such determination despite sustaining a bayonet wound to the head, was awarded the Victoria Cross.

On 11 March 1917, the British entered Baghdad but, determined to keep the pressure on the Turks, General Maude continued to advance up the twin axes of the Tigris and the Euphrates.[91] The Turks were far from defeated and fought with determination. As ever, the Gurkhas were in the thick of the fighting in the numerous battles that took place as the British advanced. By the spring of 1918, the British move along the Euphrates had reached the town of Khan al Baghdadi. The Turks occupied a strong defensive position and their commander, Nazim Bey, was expected to put up considerable resistance as his predecessor had been dismissed for not stopping the British.[92] On 26 March 1918, two brigades, the 50th and the 42nd, assaulted the position whilst a hastily assembled mobile strike force, based on the 11th Cavalry Brigade, conducted a flanking movement to get behind the defending Turks.[93] The British achieved a notable victory with some 5200 of the enemy surrendering.[94] 2/5th, 2/6th and 1/5th Gurkhas all played a key role as part of 42nd Brigade's assault on the second line of Turkish defences.[95,96] A similar tactic was employed to excellent effect at the Battle of Sharqat in late October 1918 as British forces advanced towards Mosul. 1/7th and 1/10th both took part in the battle.[97]

Soldiers from 3/3 Gurkha Rifles on sentry duty in Palestine in 1917

Further north in Palestine, General Sir Edmund Allenby, who had taken over command of the British Forces in Egypt in June 1917,[98] was also making good use of his Gurkha troops which included 1/1st, 2/3rd, 3/3rd, 2/7th, 1/8th and 4/11th.[99] The battalions distinguished themselves in numerous battles under Allenby's command with Rifleman Karanbahadur Rana, then only 19 years of age and serving in 2/3rd Gurkhas, being awarded a Victoria Cross for his actions at El Kefr on 10 April 1918. Interestingly, a detachment of 30 soldiers from 2/3rd Gurkhas also served as volunteers with T E Lawrence and his irregular army of Arabs. Along with a detachment of Indian troops, they provided mortar and machine gun teams to support his tribal forces in their fight to defeat the Turks.[100]

On 30 October 1918, the Turkish Army eventually surrendered.[101] Though the British achieved success in Mesopotamia, it came at a high cost; 30,000 of the 250,000 Allied troops which took part in the campaign died.[102] At the time, many senior officers felt that the resources that were committed to the campaigns in Mesopotamia and Palestine would have been far better used against the 'main enemy' on the Western Front.[103]

The Great War formally ended with the signing of the Treaty of Versailles on 11 November 1918. For many regiments in the British Army,

Soldiers from 2/9th Gurkhas mounting guard in Mesopotamia in 1919 after the surrender of the Turkish Army on 30 October 1918

the subsequent decades would be relatively quiet until the outbreak of the Second World War. But this was not the case for the Gurkhas; it was 'business as usual' for those policing the Empire and, as the next chapter shows, the Gurkhas were kept fully employed.

Bagpipes and Tartan in the Brigade of Gurkhas

The Gurkhas have a long history of fighting alongside soldiers from Scottish regiments going back to the Second Afghan War (1878–1881). This photograph was taken in Mesopotamia in 1917 and shows a private from the Black Watch greeting a rifleman from the 7th Gurkhas

Gurkhas have a long history of fighting alongside soldiers from Scottish Regiments. As explained in Chapter 3, Gurkhas notably fought alongside the 72nd (later the 1st Battalion The Seaforth Highlanders) during the Second Afghan War (1878–1881) and alongside the King's Own Scottish Borderers and the Gordon Highlanders at the Battle of Dargai in 1897. The relationship between the 72nd and the 5th Gurkhas was so strong and so readily apparent that, when he was raised to the peerage, General Sir Frederick Roberts, who had commanded men from both regiments during the Second Afghan War, included a highlander from the 72nd and a rifleman from the 5th Gurkhas on his coat of arms.

The Pipe Band of the 1st Gurkhas circa 1896. 1/1st Gurkhas wore the Government Mackintosh Tartan whilst 2/1st Gurkhas wore the Government Mackenzie Tartan

A piper from the 3rd Gurkhas. The photograph was taken circa 1907 in Almorah, India, which was the regiment's home location. The 3rd Gurkhas wore the Colquhoun Tartan, a tradition that continued after Indian Independence in 1947

The close relationship between Scottish soldiers and Gurkhas led to some Gurkha regiments forming pipe bands. The first of these, which was trained by the 2nd Battalion of The King's Own Scottish Borderers, was formed in 1885 by Colonel Hay of the 1st Battalion, the 4th Gurkha Rifles.[1] The 2nd Battalion of the 4th Gurkhas formed its own pipe band in 1887 at Peshawar. This band was trained by the Highland Light Infantry who gave permission for the band to wear their Mackenzie tartan.

The 1st Gurkha Rifles also formed a pipe band in 1886, followed by the 3rd Gurkhas in 1888. The other Gurkha regiments quickly followed these examples, forming their own pipe bands with the exception of the 2nd Gurkhas. This latter regiment retained a bugle band to commemorate the close relationship it had established with the 60th Rifles at the Siege of Delhi in 1857 (see Chapter 3).

Many of the Gurkha regiments adopted some of the highland dress worn by pipe bands in Scottish regiments.

The pipe and drum majors of 5/9th Gurkhas at a VE Day Parade in 1945. In 1914, the regimental pipers of the 9th Gurkhas were granted permission to wear the Duff Tartan. This was discontinued after the First World War and plain green plaids were worn instead

A piper, drummer and bugler from the 8th Gurkhas circa 1930.

The Pipe Band of the 42nd Regiment (which changed its name to the 6th Gurkhas in 1903)

The pipe and drum majors from the 10th Gurkhas saying farewell to their opposite numbers from The Royal Scots in 1960. In 1924, King George V gave authority for the 10th Gurkhas to wear the Hunting Stewart Tartan

Notably, however, Gurkhas have always favoured trews over kilts. A variety of tartans were adopted by the Gurkhas reflecting a particular relationship with a Scottish regiment or, in some cases, the family tartan of the colonel of a specific Gurkha regiment if he were a Scot. Interestingly, after Independence in 1947 the Indian officers of the 3rd Gurkhas wrote to the Clan Chief of the Colquhouns to seek his permission for their pipers to continue wearing his tartan. He gladly gave it!

Pipers from 7th Gurkhas (left) and 10th Gurkhas (right) flank Colonel C W Yeates DSO at a post Second World War reunion

The tartan worn by the pipe bands of the Gurkha infantry regiments occasionally changed over the years as regiments sought to recognise particular events. In 1949, for example, the Cameronians (Scottish Rifles) were officially affiliated with the 7th Gurkhas. To commemorate this, the Cameronians requested that the 7th Gurkhas wear their tartan (the Douglas) rather than that of the Black Watch which had been worn by the 7th Gurkhas since 1926.

Pipe bands continue to play an important role in today's Brigade of Gurkhas. In addition to making an

The pipes and drums of the 6th Gurkhas lead a battalion parade along the Hong Kong border with China on 3 July 1991, commemorating the battalion's last ever border tour. The 6th Gurkhas did not have a tartan and wore plain green plaids

invaluable contribution to regimental functions, every year they also take part in numerous national and international events, raising the profile of the Brigade of Gurkhas. A high level of musical competence is maintained by attendance on courses at the Army School of Bagpipe Music and Highland Drumming in Edinburgh.

Soldiers of The Royal Gurkha Rifles (RGR) also wear a patch of Hunting Stewart Tartan behind the RGR cap-badge on the Hat Felt Gurkha to mark an affiliation with The Royal Scots (The Royal Regiment). The affiliation was formally announced between the latter regiment and the 10th Gurkhas in 1950 and the tradition has been continued in the RGR.

In 1994, the RGR provided a Gurkha Reinforcement Company (GRC) to serve with The Royal Scots. The company, known as B (Gallipoli) Company, served with The Royal Scots (The Royal Regiment) in Bosnia as part of the Stabilisation Force (SFOR) and in 24 Airmobile Brigade. It transferred to the newly formed The Highlanders in 2001, serving in the Falkland Islands and again in Bosnia before being disbanded in 2004.

Current tartans of the Brigade of Gurkhas

The Royal Gurkha Rifles

The Queen's Gurkha Engineers

Queen's Gurkha Signals

The Queen's Own Gurkha Logistic Regiment

The pipes and drums of The Royal Gurkha Rifles performing with the Band of The Brigade of Gurkhas in Brunei in 2010. Note the trews of Douglas Tartan worn by the piper and drummer

CHAPTER 5

No Respite: the Inter-War Years

1919–1937

In 1917, the British Government agreed to the 'progressive realisation of responsible government' in India.[1] This might suggest that the British understood the imperative for India and its other colonies to begin the transition to self-rule. But this was not the case and, for the Gurkhas at least, the inter-war years were spent fighting as part of the Indian Army to maintain control of Britain's most valuable colony.

Perhaps the most infamous incident of this period which involves Gurkhas occurred in Amritsar on 13 April 1919. Unrest had been growing across India for some time with large gatherings taking place in March and April 1919. These were supposed to be demonstrations of 'passive resistance'[2] but, as the British overreacted to events, they

First aid instruction in the 1920s

became increasingly violent. Things came to a head on 10 April when Miss Marcia Sherwood, a medical missionary who had worked in Amritsar for the previous 15 years, was attacked by a mob and left for dead.[3] This so inflamed Anglo-Indian tempers that, on 11 April 1919, control of Amritsar was passed to the military and to the local commander, Brigadier-General Reginald 'Rex' Dyer.[4] General Dyer was 54 years old at the time. He had made his name in 1916 leading a force of some 150 men against 3000 tribesmen on the Baluchistan-Persian border.[5] Popular with his soldiers, he was tough and highly experienced. But he was also in constant pain as a result of wounds and injuries sustained during the course of his long career.[6] He was therefore irritable, short-tempered and, following the 'unspeakable attack' on Miss Sherwood,[7] in no mood to compromise with the 'natives'. The instructions he received were clear: 'no gatherings of persons nor processions of any sort will be allowed. All gatherings will be fired upon.'[8]

Brigadier-General Reginald 'Rex' Dyer who commanded the force at Amritsar on 13 April 1919 when 379 protesters were massacred. Of the 50 men in General Dyer's hastily assembled force, 25 were from the 9th Gurkhas

On 13 April, a crowd of some 20,000 people gathered in the Jallianwalla Bagh, a meeting ground in Amritsar surrounded on all sides by high walls.[9] The gathering was a clear breach of the rules and Dyer did not hesitate,[10] personally leading a detachment of 25 men from the 9th Gurkhas and a mixed force of 25 men drawn from the 54th Sikhs

The Jallianwalla Bagh where, on 13 April 1919, the Amritsar Massacre took place

and 59th Scinde Rifles to impose order.[11] As soon as his troops were deployed, he gave the order to open fire. No warning was given and the crowd was unable to disperse given the confined nature of the ground.[12] After ten minutes of sustained shooting, 379 demonstrators lay dead and more than 1500 were wounded.[13] Reaction to the event was mixed. Although Winston Churchill subsequently described the massacre as 'monstrous',[14] some, at least initially, believed it was justified given the 'spectre' of the 1857 Mutiny and the need to impose a 'moral lesson' on those who would seek to overthrow Britain's rule.[15] Interestingly, on his deathbed in 1927, Dyer reportedly said:

> so many people who knew the condition of Amritsar say I did right ... but so many others say I did wrong. I only want to die and know from my Maker whether I did right or wrong.[16]

Across the border in Afghanistan, the then amir, Amir Aminullah Khan, sensed an opportunity to exploit what he perceived to be Britain's inability to maintain control in India. He made a declaration of Jihad (or 'Holy War') against the British, claiming that they had inflicted all kinds of injustices against 'religion, honour and modesty.'[17] On 3 May 1919, a detachment of 150 Afghans crossed the border at the western end of the Khyber Pass. They quickly seized the high ground surrounding the British fort at Landi Kotal, as well as the village of Bagh which

A distant photograph of the British fort at Landi Kotal. One of the fort's chief vulnerabilities was its reliance on the village of Bagh for its water source

controlled the fort's water supply.[18] The British reacted by declaring war on Afghanistan on 6 May 1919, beginning the Third Anglo-Afghan War.[19] A force, which included 2/1st, 1/11th and 2/11th Gurkhas, was quickly despatched by lorry to relieve the besieged garrison at Landi

The Khyber Pass in the early 1920s. The Pass was strategically important as it was one of the few routes from Afghanistan into British India. It was the scene of considerable fighting during the Third Anglo-Afghan War of 1919

Kotal.[20] After two significant engagements, known as the First and Second Battles of Bagh, the Afghans eventually withdrew. Notably, 1/11th played a key role in the Second Battle of Bagh on 11 May 1919, smashing through the Afghan defences to take the centre of the village.[21] The withdrawing Afghans had a difficult time. Not only were they strafed by British aircraft but they were also ambushed by cut-off groups which the British, now expert at frontier warfare, had positioned on the likely escape routes.[22] The Afghan casualties were significant. In the Second Battle of Bagh alone they lost 100 killed and 300 wounded compared to the British casualties of eight killed and 29 wounded.[23]

Recognising the tactical importance of the Khyber Pass as a route into India, the British despatched a brigade to secure it. By 13 May 1919, the brigade had advanced across the border and occupied the town of Dacca. Although this was achieved without opposition, the Afghans soon brought up reinforcements. The British encampment near the town was poorly sited and the Afghans were able to bring artillery and small arms fire to bear, inflicting considerable casualties.[24] Two battalions, the

Gurkhas guarding captured tribesmen in Waziristan circa 1920

1/35th Sikhs and the 1/9th Gurkhas,[25] were therefore sent to clear the surrounding hills but the attack faltered until further reinforcements, which included 2/1st Gurkhas, were despatched to support them.[26]

The situation on the North West Frontier was becoming increasingly precarious. Not only were the British having to fight the regular Afghan Army which, at the time, comprised some 78 infantry battalions, 21 cavalry regiments and 280 breech loading artillery pieces,[27] but they were also having to contend with border tribesmen who, incited by the call to Jihad, were taking every opportunity to make life difficult for the British.

In the two previous Anglo-Afghan wars, the British had outflanked the Afghans by taking the southern city of Kandahar.[28] Determined to avoid a repeat, Amir Aminullah Khan had deployed a large number of troops to reinforce the fortress at Spin Boldak, an Afghan town that straddled the main route from the border to Kandahar.[29] Notwithstanding this, the British decided to cross the border and advance towards Kandahar in order to draw reserves away from the fighting in the north and to stop the Afghans from inciting rebellion amongst the local tribesmen.[30] The British force, which was supported by aircraft, consisted of four infantry battalions, two of which were Gurkha,[31] one machine gun company and 12 artillery pieces.[32] The force moved quickly but, whilst the British achieved surprise,[33] it was some time before a detachment of Gurkhas was able to scale the walls and breach the fort's defences.[34] The defenders put up a spirited resistance, fighting to the bitter end. Of the 600 Afghans defending the fortress, over 170 were killed and 176 taken prisoner.[35] The remainder were either wounded or managed to escape.

The Battle of Spin Boldak which took place on 27 May 1919. The Afghan defenders were taken by surprise and, after much hand-to-hand fighting with kukri and bayonet, the fort was captured

The main Afghan offensive took place south of the Khyber Pass and north of Kandahar in central Waziristan.[36] The Afghan's intention was to use overwhelming force, pushing across the border on two axes with a force comprised of 14 infantry battalions and 48 guns.[37] By late May 1919, the Afghan Army had advanced as far as Thal, an important railway junction on the main line to Peshawar. The British garrison at Thal, which included 3/9th Gurkhas,[38] was soon surrounded. The Afghan

Lieutenant Colonel Alexander Charles Broughton who commanded 3/9th Gurkhas during the Third Afghan War

artillery was quickly brought into action and, on the first day, inflicted considerable damage to ration trucks, a petrol dump, stores of animal fodder and a wireless transmitting station.[39] They also succeeded in capturing the water pumping station that supplied the garrison.[40] The Afghans appeared to have the tactical advantage but the commander, Nadir Khan, failed to exploit it. Believing that the tribes to the east of Thal would prevent any British relief column from reaching the town, he preferred to lay siege to the garrison rather than risk the heavy casualties that he knew would result from a frontal attack.[41] It would prove to be a fatal mistake.

Fearing the loss of Thal, the British assembled a relief force under the command of Brigadier-General 'Rex' Dyer. His actions at Amritsar might have shown a lack of judgement but he was no fool and he set out to relieve Thal by deceiving the Afghans. Using logs as dummy artillery pieces and tying brush to the mud flaps of his vehicles to throw up dust clouds, he tried to create the impression that a large force was on the move to rescue the garrison at Thal.[42] Nadir Khan allegedly fell for the ruse, stating 'in the name of God, we have the whole artillery of India coming against us.'[43] Dyer arrived at Thal on 31 May 1919 having covered the last 18 miles in just 12 hours,[44] a remarkable achievement given the terrain.

A signaller from the 5th Gurkhas circa 1919. The heliograph was an effective means of communicating over large distances in the mountains of the North West Frontier

The Afghan defences comprised of two wings held by irregulars with, in the centre, a force of some 19,000 regular troops on the hills to the west of the town.[45] Dyer again proved his competence as a field commander; using this artillery, he feigned an attack on the northern wing whilst launching his ground troops against the southern one.[46] After several hours of fighting, the Afghans on the southern flank dispersed.[47] Hoping to buy time to withdraw his troops, Nadir Khan sent a message to Dyer requesting that he observe an armistice that had been agreed whilst negotiations between the two forces' respective governments were underway.[48] Suspecting a ruse, Dyer's response was that 'my guns will give an immediate reply.'[49] The next morning on 2 June 1919, Dyer launched his force, which included the 3/9th Gurkhas from the now liberated Thal,[50] against Nadir Khan's main defensive position. The attack, which was preceded by an artillery bombardment, was far easier than expected, largely because the bulk of Nadir Khan's 19,000 regular troops had already withdrawn and were three miles away when the bombardment began.[51] Although the battle at Thal was the last major incident of the Third Anglo-Afghan War, the war did not formally end until 8 August 1919 when, after two months of negotiations, both parties signed the Treaty of Rawalpindi.

The British had succeeded in driving the Afghan army back across the border but it was soon apparent that the rebellious tribesmen of Waziristan, whipped into religious fervour by the call to Jihad, were going to be just as difficult to pacify. Of the tribes in Waziristan, the Mahsud were the most troublesome and made it very clear that they had no intention of recognising British authority.[52] They knew that they couldn't win against the might of the British Empire but they hoped to make victory so expensive for the British that they would be left alone.[53] By the end of November 1919, the British finally had enough

A supply column on the move in Waziristan circa 1919/1920. Although roads were subsequently built to aid movement, camels and mules would remain the primary means of transportation throughout the inhospitable terrain for many years

Mahsud tribesmen bringing in wood circa 1920 near Kaniguram. Tough and resourceful, the Mahsud were formidable opponents.

of the Mahsud's behaviour.[54] They assembled a force, known as the Derajat Column, to advance into Mahsud territory and to impose discipline. The force consisted of the 43rd and 67th Infantry Brigades along with three batteries of mountain artillery and other supporting arms.[55] The two infantry brigades included four Gurkha battalions (4/3rd, 2/5th, 2/9th and 3/11th).[56] The initial advance into Mahsud territory went reasonably well and, by 13 December 1919, the force had established a temporary camp a few miles north of the Mahsud village of Jandola.[57] The next phase of the plan was to strike deep into the Mahsud heartland to seize the fortified villages of Makin and Kaniguram.[58]

A Mahsud tribesman. Adept at concealment and ambush, the Mahsud were excellent shots. Their preferred technique was to snipe at British forces from a distance and then disappear into the mountains before a counter-attack could be organised

Within a few weeks, the Derajat Column would be involved in what has been described as 'the fiercest battle in Frontier history'.[59] To understand why, we need to look at the tribesmen opposing the British in a little more detail. The Mahsud were skilled and experienced guerrilla fighters. Able marksmen and adept at concealment, many had served in militias or regular units during the Great War. They therefore had a sound understanding of British tactics, particularly the principle of 'fire and manoeuvre', which they employed to excellent effect.[60] Following the Great War there was also a surplus of modern high velocity weapons and the Mahsud tribesmen had little difficulty replacing their aging Martinis with more accurate and effective rifles. Critically, these didn't create a puff of smoke when they were fired.[61] This was a significant improvement as the British had become adept at using the smoke to identify the sniper's position. The commander of the Derajat Column, General Sir Andrew Skene, was a veteran of frontier warfare and was well aware of how capable the Mahsud were. He had explained his concerns to the British Government but he had been told just to 'get on with the job.'[62]

The only practicable way to reach the villages of Makin and Kaniguram was up the narrow and steeply sided valley of the Tank Zam River.[63] Before he could advance, Skene first needed to picket the high ground on either side of the valley. This was a formidable challenge, particularly as the Mahsud knew the area intimately. On 19 Dec 1919, Skene deployed several battalions to try and secure the high ground that dominated the southern approach to the valley. It was a desperate business with one battalion, the 103rd Mahrattas, being chased down the mountainside, across the Tank Zam and back to their camp. The battalion lost 95 killed, including their commanding officer, and a further 140 wounded.[64] A second attempt to secure the high ground was made on the 20 December 1919. Although the infantry succeeded in establishing a defensive position on a high bluff that overlooked the

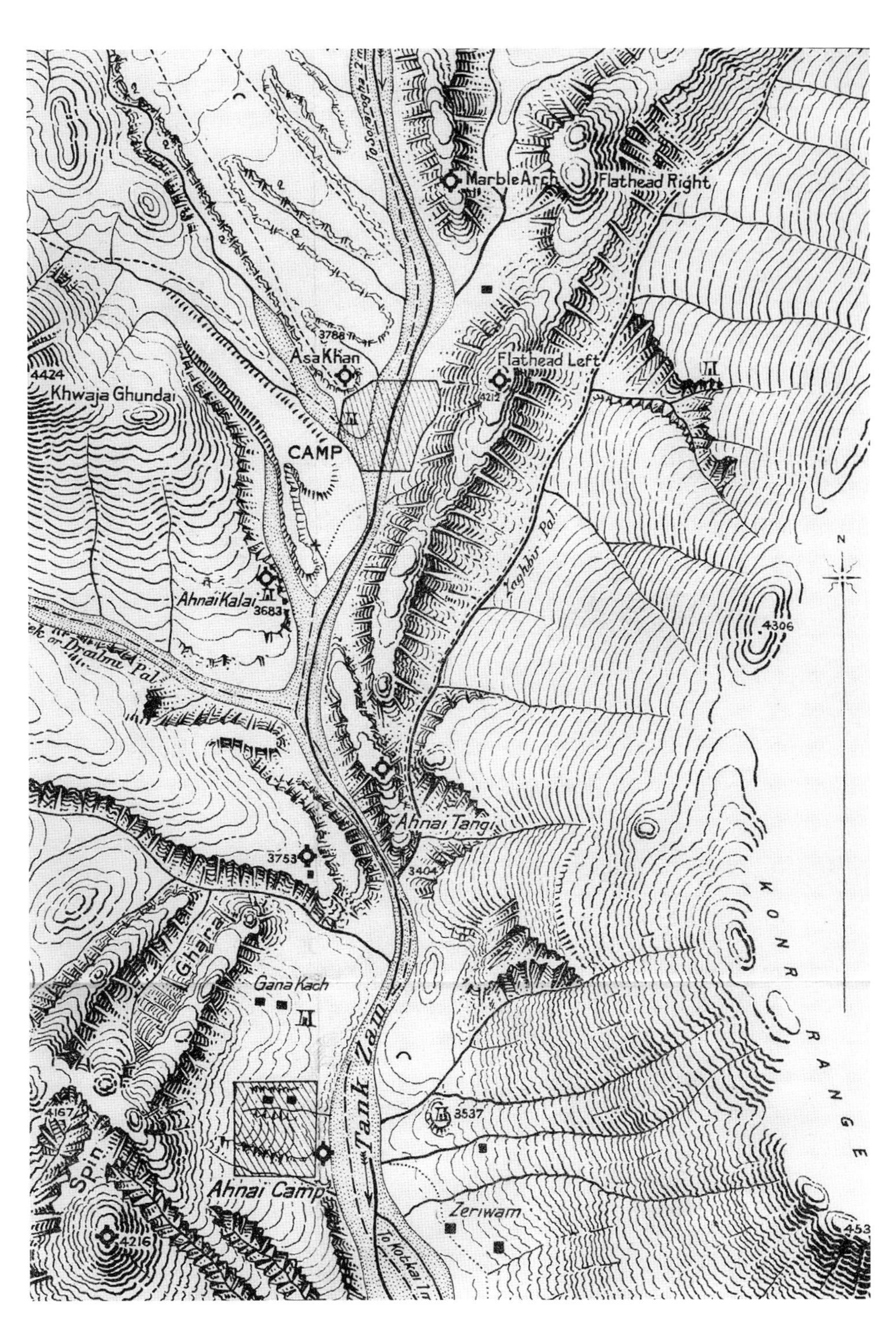

Map showing the severity of the terrain either side of the Tank Zam River. The Derajat Column advanced from the south and encountered stiff resistance as it forced its way through the Ahnai Tangi Gorge

river, the position was attacked and over-run during the night.[65] A third attempt to secure a dominating position, this time on the north side of the river, achieved a degree of success and, after much bloody hand-to-hand fighting, the column was able to advance up the valley as far as a feature known as the Ahnai Tangi.

The Ahnai Tangi was a truly formidable obstacle. A gorge with sheer walls rising over 300 ft, the Tank Zam became a raging torrent as it forced its way through the 30 ft gap separating the walls of the gorge.[66] The plan to take the Ahnai Tangi was straightforward. 2/5th Gurkhas would secure the right flank, on the eastern side of the river, with 2/9th Gurkhas securing the less challenging left flank, on the western side of the river.[67] On 14 January 1920, 2/5th Gurkhas advanced along the ridge until they reached a feature known as Flathead Left.[68] Although high, it was overlooked by an even higher feature half a mile to the north east known as Flathead Right. This was occupied by a large number of Mahsud who were able to direct a steady stream of lethal fire at the Gurkhas, inflicting heavy casualties.[69]

Numerous attacks were repulsed by the Gurkhas who, running low on ammunition, made good use of their kukris. The situation became increasingly desperate with the battalion's commanding officer, Lieutenant Colonel Crowdy, being killed leading a bayonet charge against the attacking tribesmen.[70] Eventually, the 2/76th Punjabis were sent to

Gurkhas of the Derajat Column engaged in hand-to-hand combat with Mahsud tribesmen as they fight their way through the Ahnai Tangi in January 1920

The officers of the 5th Gurkhas (Frontier Force) taken in 1908. J D Crowdy, seated on the ground second from the left, would be killed 12 years later leading a bayonet charge against assaulting Mahsud tribesmen as the Commanding Officer of 2/5th Gurkhas

reinforce the Gurkhas.[71] They passed through the Flathead Left position and continued down into the couloir which separated the two features with the aim of seizing Flathead Right. But the Mahsud were waiting and, having sustained 60 casualties in the first 100 yards, the Punjabis withdrew back to Flathead Left.[72] Additional troops were then sent to reinforce the Gurkhas' position, securing it against further attacks. With the Flathead Left feature secured, the remainder of Skene's force was able to pass through the Ahnai Tangi and on towards the villages of Makin and Kaniguram. Once there, they punished the Mahsud tribesmen by tearing down their fortified towers, destroying the terracing in their fields and breaking the irrigation channels.[73] The Mahsud eventually sued for peace but it had come at a cost: of the men that made up the Derajat Column, 639 were killed during the campaign and a further 1683 wounded.[74]

Having suppressed the Mahsud, the decision was taken to occupy Waziristan. Roads and forts were therefore built to support a more permanent presence, enabling troops to be moved quickly from one hot spot to another. The Gurkha battalions settled into the routine of fighting on the North West Frontier and then returning to their home base locations to rest, recover and train for their next deployment. Occasionally, however, this routine would be broken as one group of tribesmen or another committed an outrage that demanded a more deliberate response. One such incident happened in the spring of 1937. Mirza Ali Khan, who became known as the Fakir of Ipi, had been stirring up unrest amongst the Pashtun population in Waziristan for several years.[75] The Ipi claimed that the roads and garrisons that had been established throughout Waziristan were '... but a prelude to the full occupation of the tribal areas, the destruction of Islam and the expropriation of their lands.'[76] He declared a Jihad and, by early 1937, the British found themselves engaged in a low intensity campaign against the fakir and his followers.[77] In order to secure their transit routes, the British established permanent pickets on the high ground dominating key roads. Though effective, they were ideal targets for the tribesmen who, ever keen to avoid direct force-on-force engagements against the superior firepower of the British, would mount intermittent attacks on the isolated outposts.

The British constructed small defensive positions, known as sangars, on high ground dominating the many roads that they built throughout Waziristan

A contemporaneous montage showing the Gurkhas' defensive position (or sangar) at Damdil and the six survivors of a series of determined attacks by Pathan tribesmen on 20 March 1937. Top row from right: Rifleman Dhanbahadur Gurung IOM; Naik Corporal Nandalal Ghale IOM; Rifleman Nandabir Thapa IDSM. Bottom row: Rifleman Ganjasing Gurung IDSM; Rifleman Khumansing Gurung IDSM; Rifleman Dholi Gurung IDSM. Rifleman Dalbahadur Gurung and Rifleman Uttamsing Rana were both killed defending the position

On 20 March 1937, a determined attack was made against a picket at Damdil manned by a detachment of eight soldiers from 2/5th Gurkhas.[78] Fierce fighting continued throughout the night with the tribesmen mounting numerous assaults. Armed with swords, rifles and grenades, they were a formidable force but the Gurkhas held firm until, as dawn broke, the tribesmen were driven off by a relieving force of armoured cars, artillery and aircraft.[79] Of the eight Gurkha defenders, two were killed and the remainder seriously wounded.[80]

The campaign against the Fakir of Ipi and his followers continued throughout the remainder of the year. By November 1937, his support had dwindled to such an extent that he was unable to maintain any credible resistance against the British presence.[81] Although the fakir's supporters numbered no more than a few thousand tribesmen, 61,000 troops had been required to suppress the insurgency.[82] Of these, 245 had been killed and a further 684 wounded.[83]

The counter insurgency skills that the Gurkhas had learnt on the North West Frontier would stand them in good stead in later years in Borneo

Above 1/9th Gurkhas on the march in Waziristan in 1935. Roads were constructed throughout Waziristan to enable troops to move quickly from one hot spot to another

Below The severe nature of the terrain meant that camel borne stretchers were often the most effective way of recovering the wounded following engagements with tribesmen on the North West Frontier in Waziristan

A soldier from 9th Gurkhas having a field haircut, Waziristan circa 1930

and Malaya. But, before then, the world would once more descend into chaos as the ambitions of an expansionist Germany initiated the Second World War. Perhaps inevitably, the Gurkhas would have a key role to play.

The dining room in the officers' mess of the 9th Gurkhas circa 1920 in their regimental home at Dehra Dun

Recruiting in the Brigade of Gurkhas

A military doctor examining a potential new Gurkha recruit in 1902. Then, as now, the standard for joining Britain's Brigade of Gurkhas was particularly high, though some requirements, such as the minimum height, were relaxed during the First and Second World Wars

Potential recruits at the start of the selection process in 1941

Gurkha soldiers continue to be recruited from the male population of Nepal under the terms and conditions of the Tri-Partite Agreement signed between Britain, Nepal and India in November 1947. Every year, the Brigade of Gurkhas deploys teams into the hills of Nepal to conduct the initial phases of its recruit selection process. This begins with an advertising phase which runs from April to May each year. In 2014, Senior Recruit Assistants (SRAs – formally known as '*galah wallas*') travelled to 56 of Nepal's 75 districts during this phase, briefing potential recruits as well as key people in the local community. During the briefings, the SRAs explain the selection procedure and the standards that candidates have to achieve in the various assessments, dispelling many of the myths that exist, particularly in the more remote areas.

The advertising phase is followed by a registration phase which takes place from May to July each year. During registration, recruiting assistants carry out a basic assessment of each potential recruit to ensure that he meets the selection criteria. Some of the current entry criteria are:

- **Age**. Must be between 17 years and six months and 21 years as of January of the year in which selection is taking place.
- **Height**. Not less than 158cm (just over 5 feet and 2 inches).
- **Weight**. Not less than 50kg (just over 7 stone and 12 lbs)
- **Chest**. Not less than 79cm (just over 31 inches).
- **Education**. Must have completed the Nepalese School Leaving Certificate in at least the 3rd Division (which approximately equates to having achieved three 'C' grades and two 'D' grades at GCSE).

During the initial phase of recruit selection, Senior Recruit Assistants (SRAs) travel to towns and villages throughout Nepal to brief potential recruits, as well as members of the community, on the standards required for service in the British Army. This advertising phase is then followed by a registration phase. Potential recruits that meet the criteria during the registration phase are subsequently called forward to Regional Selection in Pokhara and Dharan

If an individual meets the required standard, he is called forward to Regional Selection. This takes place between July and September each year at two locations in Nepal, Pokhara in the west and Dharan in the east. The selection process lasts a day and includes an educational assessment, a medical examination, physical tests and interviews with both a Nepali Gurkha officer and a British Gurkha officer. Individuals are placed in an order of merit based on their performance. The final phase is Central Selection which lasts three

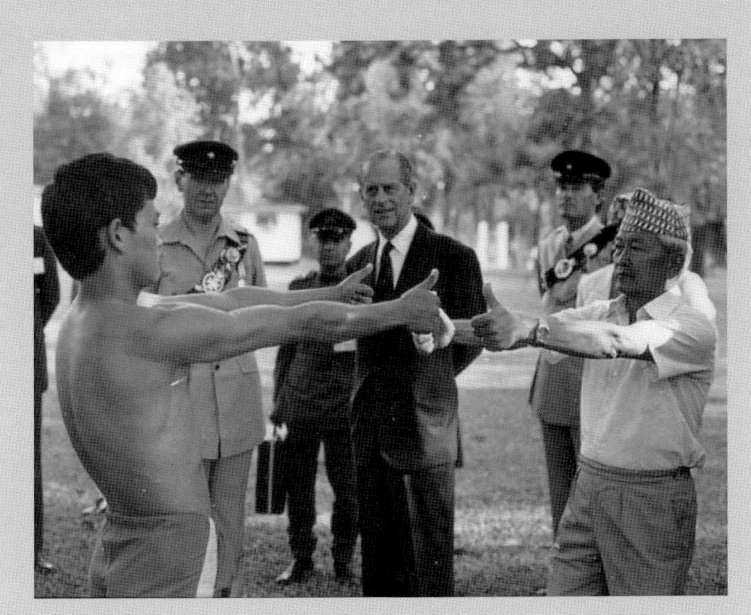

His Royal Highness The Duke of Edinburgh observing recruit selection in 1986. The Duke of Edinburgh was the Colonel-in-Chief of the 7th Duke of Edinburgh's Own Gurkha Rifles

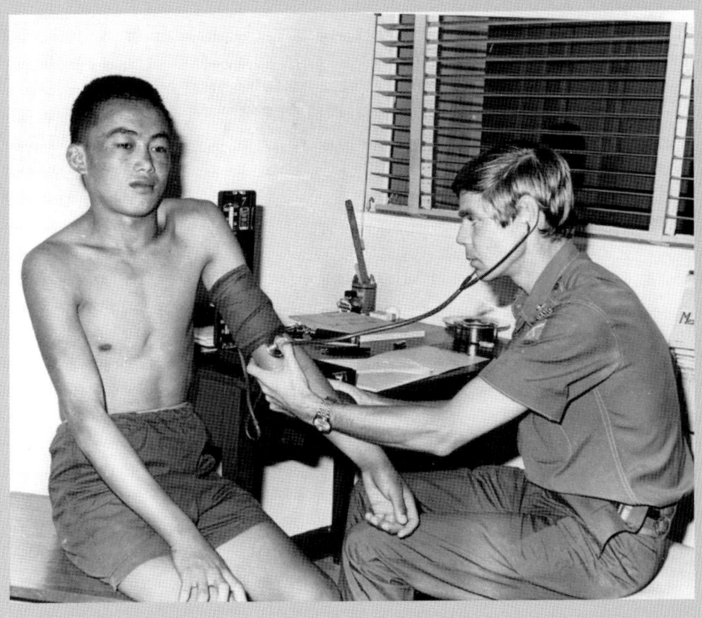

A potential recruit undergoes a medical during the selection process in 1973

Potential recruits circa 1920. The transition from 'hill boy' to Gurkha soldier begins with the individual having the courage to put himself forward for initial registration which takes place across Nepal from May to July each year

Potential recruits for the 5th Gurkhas, taken during the Second World War

weeks and takes place in the British Army's camp in Pokhara. Only the top 500 potential recruits from Regional Selection are called forward to take part in Central Selection.

The standards that potential Gurkha recruits have to achieve during recruit selection are particularly demanding. The Brigade of Gurkhas can set the highest of entry standards because there are many more applicants than places. In 2014, for example, there were 7865 applicants for 230 places, an increase of 3.9 per cent over the previous year.

The current physical standards that potential recruits have to achieve are:

- **800 metres run.** Run 800 metres in not more than 2 minutes 40 seconds.
- **Heaves/Pull-Ups.** Perform not less than 12 underarm heaves/pull-ups.
- **Sit-Ups.** Perform not less than 70 sit-ups in 2 minutes on flat ground.
- **1.5 Mile/2.4km Run.** Run 1.5 miles in not more than 9 minutes 40 seconds.
- **Stamina Assessment.** Complete a 5km route (with a height gain of approximately 450 metres) carrying 25kg in not more than 48 minutes.

The stamina assessment is particularly challenging. Although the course is only 5km in length, the height gain and the weight that potential recruits have to carry make this a formidable test. The weight has to

Potential recruits warming up with an instructor prior to completing one of the physical assessments during recruit selection

Potential recruits have to be able to complete at least 12 under-arm heaves/pull ups. This is not at all easy and the majority of young men will have practised this for a number of months prior to attending recruit selection

Potential recruits have to complete at least 70 sit-ups in 2 minutes during the recruit selection process

Potential recruits have to complete a number of physical tests during the selection process. One of these is a 1.5 mile run which has to be completed in 9 minutes 40 seconds. Numbers are painted on the recruits' chests to enable them to be identified

be carried using the 'doko', a rattan basket that the hill people of Nepal traditionally use to carry loads along the mountain paths of the Himalayas.

Gurkha units have to operate within the wider context of the British Army. The working language within the Brigade of Gurkha is therefore English. Because of this, recruits are expected to have a reasonable grasp of the language before they join the British Army. Standards are assessed during Central Selection by means of an essay as well as reading and listening comprehension exercises.

At the end of the three week Central Selection, the 500 potential recruits are again placed in an order of merit. Only the top 230 will be offered places in the British Army. Once they have accepted a place, they will be invited to swear allegiance to the British Crown,

A potential recruit taking part in the stamina assessment conducted during Central Selection in Pokhara, West Nepal. During the assessment, which is conducted as a race, potential recruits have to carry a weight of 25kg in a 'doko', the rattan basket that the people of Nepal traditionally use to carry heavy loads along the mountain paths of the Himalaya

Potential recruits waiting to start the stamina assessment. Although the course is only 5km in length, it starts in the bottom of a river valley near the British Camp in Pokhara and climbs over 450 metres to the high ground shown in the picture. The footbridge in the background spanning the valley is typical of footbridges in Nepal. Traditionally made of hemp rope, these are often now made of steel wire. Notwithstanding this, crossing them, particularly at night and in bad weather, is not for the faint-hearted!

During Central Selection in the British Army camp in Pokhara, potential recruits have to write essays and complete both reading and listening comprehension assessments

a process known as 'attestation', before being flown to the UK to begin a nine month training package at the Infantry Training Centre (ITC) in Catterick, North Yorkshire.

The Brigade of Gurkhas aims to recruit 50 per cent of its soldiers from the east of Nepal and 50 per cent from the west. This mix has historical origins. Of the four Gurkha Infantry Regiments that transferred to the British Army in 1947, the 2nd and 6th Gurkhas were recruited from the west and the 7th and 10th from the east. Although some transfers between the infantry battalions do take place, the current battalions of The Royal Gurkha Rifles perpetuate this geographic division with the soldiers in 1RGR being predominantly from the

New recruits of the 3rd Gurkhas swearing their allegiance to the British Crown in 1944. This is known as attestation. Having 'attested', each recruit is given a kukri, the traditional fighting knife of the Gurkhas

Recruits are taught to use the kukri during the training they receive at the Infantry Training Centre in Catterick

The spiritual needs of Gurkhas are well catered for in the British Army. All Gurkha units have a Hindu 'Pundit' permanently attached and Buddhist monks visit each unit regularly. This picture shows a Buddhist monk carrying out a religious ceremony for recruits

west whilst those in 2RGR are predominantly from the east. The supporting arms are completely mixed with all squadrons having approximately the same number of Gurkhas from the east and the west of Nepal.

All Gurkhas are trained as infantry soldiers during the nine month training package at the Infantry Training Centre in Catterick, regardless of whether they will eventually serve in the infantry or not. This 'band of brothers' approach ensures that all Gurkhas have a basic understanding of infantry tactics, widening their utility.

Like the majority of other regiments in the British Army, Gurkha units are officered by a mix of late entry officers (soldiers who have been promoted through the ranks and then commissioned) and direct entry officers (young men and women who have been specifically recruited as officers). Direct entry officers complete a nine-month training programme at the Royal Military Academy Sandhurst before being commissioned. After Sandhurst, direct entry officers complete their special-to-arm training, giving them the professional skills they need to be effective in their chosen arm or service. Officers going to The Royal Gurkha Rifles join their battalion immediately after their special-to-arm training. Officers going to The Queen's Gurkha Engineers, Queen's Gurkha Signals and The Queen's Own Gurkha Logistic Regiment usually complete a tour with a non-Gurkha unit before being posted to a Gurkha one. All officers serving with the Gurkhas are expected to speak Nepali. The Brigade of Gurkhas has a language school in Nepal which all officers attend before completing a month long trek. The aim of the trek is to enable officers to consolidate their new-found language skills as well as to improve their understanding of the environment from which their soldiers are recruited.

From recruit to trained Gurkha soldier takes nine months of hard work. Notably, The Brigade of Gurkhas maintains the British Army's highest pass rate during basic training, a tribute to the quality of the young men who join Britain's Gurkhas

Taken in 1911, this photograph shows veterans from the Indian Mutiny and the Siege of Delhi of 1857. Despite their age, the Gurkha officers in the centre of the picture still exhibit the pride and discipline that was instilled into them during their basic training decades beforehand

CHAPTER 6

The Second World War

1938–1945

On 3 September 1939, Britain declared war against Germany. But for the Gurkha battalions in India, it was largely 'business as usual.' The prevailing view in Whitehall appeared to be that the war would be 'won by the naval blockade and the trusty French: there was little need for the British Army to exert itself, none for the Indian Army to do so.'[1] Gurkha battalions continued to conduct operations on the North West Frontier but, as the quick victory over the Germans failed to materialise, brigades and divisions from the Indian Army were committed to the fight.

By the end of May 1941, the 10th Indian Division, commanded by Major General Bill Slim, had arrived in the Middle East. The division, which

8th Gurkhas on operations on the North West Frontier in 1939. For the Gurkha units of the Indian Army, it was 'business as usual' for the first few years of the Second World War. But in early 1941 this was to change as the 10th Indian Division was committed to operations in the Middle East

contained four Gurkha battalions (1/2nd, 2/4th, 2/7th and 2/10th), was initially based in Basra in southern Iraq and had the task of protecting Britain's oil supplies.[2] It was the Gurkhas' first deployment of the Second World War but it was far from their last.

On 7 February 1941, Lieutenant General Erwin Rommel arrived in North Africa.[3] It marked the beginning of a series of defeats for the Allied forces in which the Gurkha units of 10th Indian Division were to be involved. On 26 May 1942, Rommel began offensive operations to take the fortified coastal city of Tobruk.[4] The city was of strategic importance to the Allies and had to be defended at all costs. The approaches to the city were protected by six brigade-sized defensive boxes which, with minefields to their front, were designed to delay any German advances until they could be destroyed by British armour.[5] Rommel's initial assaults were repulsed by the defending forces but, on 1 June 1942, his forces eventually succeeded in breaking through the defensive belt. The 10th Indian Brigade, which included 2/4th Gurkhas, was re-deployed to try and halt Rommel's advance. Supported by artillery, the Gurkhas dug in and prepared to engage Rommel's tanks. Unfortunately for the Gurkhas, the battalions either side were quickly overwhelmed by the Germans. Surrounded on all sides, the battalion fought valiantly but the outcome was inevitable and, on 6 June 1942, the remnants of the battalion surrendered to the Germans.[6]

Field Marshal Viscount Slim. Commissioned into the Royal Warwickshire Regiment in August 1914, he fought alongside Gurkhas at Gallipoli before formally transferring to the Indian Army in 1919. At the outbreak of the Second World War, Slim was given command of 10th Indian Infantry Brigade and took part in the East African Campaign to liberate Ethiopia from the Italians. He was promoted to take over command of 10th Indian Division in May 1941 when its then commander, Major General William Fraser, was forced to relinquish command due to ill health

Gurkhas manning 6-pounder anti-tank guns during the North Africa Campaign.

Soldiers from 2/7th Gurkhas arriving at the British Lines having completed a 36 day trek through the desert after the fall of Tobruk. The battalion had been involved in the defence of Tobruk but, surrounded and running out of ammunition, it had had little option but to surrender to the attacking forces of Rommel's Afrika Corps

By 19 June 1942, Rommel's forces had advanced as far as the perimeter of Tobruk.[7] 11th Indian Brigade, which included 2/7th Gurkhas, were part of the defending force. The Germans swung round to the east of Tobruk and attacked the 11th Brigade area, concentrating their forces on a narrow front in order to punch through the Allied defences.[8] The 2/5th Mahrattas bore the brunt of the assault with 2/7th Gurkhas, who were off to a flank, helpless to support them.[9] The Germans rolled up to the Mahrattas' position, breaking through into the centre of Tobruk. 2/7th Gurkhas, surrounded and running out of ammunition, fought courageously repulsing numerous attacks but, recognising their desperate predicament, they were eventually forced to surrender.[10]

Rommel's success at Tobruk was a major victory for the Axis forces. As well as capturing the city and its garrison of some 32,000, they also seized huge quantities of fuel and supplies.[11] Rommel's advance continued east into Egypt until, in late July 1942, he was halted by General Sir Claude Auchinleck and the British Eighth Army at El Alamein.[12] At the time Auchinleck was both the overall commander-in-chief of the Middle East and the commander of the Eighth Army having sacked its previous commander, Major General Neil Ritchie, on 25 June 1942.

On 13 August 1942, Lieutenant General Bernard Montgomery assumed command of the Eighth Army. 1/2nd Gurkhas, who formed part of 4th Indian Division, soon found themselves supporting Montgomery's drive to push Rommel's forces out of Egypt. Whilst not directly involved in

Lieutenant General Bernard Montgomery, the commander of the Eighth Army, examining a Gurkha kukri. The officer looking on is Acting Lieutenant General Horrocks who commanded the Eighth Army's XIII Corps and X Corps during the campaign in North Africa

the second battle of El Alamein, they had a key role to play in fixing the German forces to the south of Montgomery's main offensive.[13]

Following their defeat at El Alamein, the Germans were slowly pushed back across the north of Africa and into Tunisia. Although the Afrika Korps was withdrawing, it fought hard, occupying a succession of

Kukris drawn, the picture, taken during the war, recreates the advance of men from the 2nd Gurkhas up a steep mountain hillside towards German defensive positions. Withdrawing west across North Africa after their defeat at the second battle of El Alamein in November 1942, Rommel's Afrika Korps occupied a series of defensive positions which had to be cleared by the advancing Eighth Army. One of the most significant of these defensive positions was the Mareth Line, originally constructed by the French to defend Tunisia against a possible attack from the east.

Gurkhas going into the assault in North Africa with kukris drawn. This picture was taken during the war but was most likely staged

Major General Francis 'Gertie' Tuker, the inspirational commander of the 4th Indian Division. Tuker persuaded the commander of the Eighth Army, Lieutenant General Bernard Montgomery, that the Fatnassa Massif was the key to unlocking the German defensive position at Wadi Akrit in northern Tunisia in April 1943. A veteran of Gurkha service on the North West Frontier, Tuker knew his Gurkhas were the right men to assault the mountain defences of the Fatnassa

strongly defended positions which the Allies had to clear. One of these was at a place called Wadi Akarit, a 15 mile gap between the sea and impassable salt marches.[14] There were two ways to take Wadi Akarit. The Allies could either force their way along the coastal strip or hook round behind the main defences by going inland and through the 'seemingly impassable' mountain massif of Fatnassa.[15] This has been described as '... a fantastic pile, like a fairy-tale mountain, split by chimneys and fissures, layered by escarpments and crowned by rock pinnacles.'[16] Montgomery favoured the coastal plain for the main assault, with a shallow hook through the mountains.

Major General Francis 'Gertie' Tuker, commanding the 4th Indian Division, believed Montgomery intended to do exactly what the Germans wanted.[17] A highly capable commander who '... combined an addiction to polo with an intellectual approach to war and a keen delight in questioning accepted military practices,'[18] he raised his concerns, arguing that the key to the German defences was the Fatnassa Massif itself as it dominated the whole position.[19] He pushed for his Division, which was experienced in mountain warfare, to be given the task of seizing this vital feature, allowing the other two divisions (the 50th Division and the 51st Highland Division) to assault the German's main defensive position on the coastal plain.[20] Montgomery accepted Tuker's proposal.[21]

After a three-mile night approach march, the lead elements of Tuker's force, which included 1/2nd and 1/9th Gurkhas, were in position to begin their assault on the enemy's defences around the summit of Fatnassa.

A painting by the artist Harry Sheldon of the action on the Fatnassa Massif, Tunisia in which Subedar Lalbahadur Thapa of 1/2nd Gurkhas earned the Victoria Cross. Capturing the Fatnassa feature was essential to enable Montgomery's Eighth Army to wheel round behind Rommel's forces[25]

Subedar Lalbahadur Thapa of 1/2nd Gurkhas who was awarded the Victoria Cross for his courageous actions at Wadi Akarit in Tunisia in April 1942. Interestingly, Lalbahadur was actually recommended for the immediate award of a Military Cross but this was upgraded to the Victoria Cross by the Army commander[28]

At 2330hrs on 5 April 1943, Tuker's forces crept silently forward.[22] They were spotted by a sentry who, though quickly despatched, somehow had time to raise the alarm.[23] The men of 1/2nd moved forward, engaging the defenders in hand to hand combat. Subedar Lalbahadur Thapa, commanding the lead platoon of D Company, raced forward to secure a narrow ravine which led up to the centre of the enemy position. Fighting his way through the defenders, he took out one machine gun post before then engaging the crew of another at the top of the ravine with his kukri.[24] With a route now clear, the rest of the company surged through onto the enemy position, quickly seizing it.

Lalbahadur's remarkable actions unlocked the enemy's position. Recognising this, the defenders launched numerous counter-attacks over the next 24 hours but the determined soldiers of 1/2nd held firm. With his flank becoming untenable and struggling under the weight of Montgomery's main assault on the coastal plain, the German commander had little option to fall back deeper into Tunisia. Subedar Lalbahadur Thapa's actions have been described as '... one of those individual feats of arms which, very rarely, tip the scales of battle.'[26] It is therefore perhaps not surprising that he was awarded the Victoria Cross, the first of 12 that would be won by Gurkhas during the Second World War.[27]

The final destruction of the Afrika Corps was left to an Anglo-American force under the command of General Dwight D Eisenhower which, in

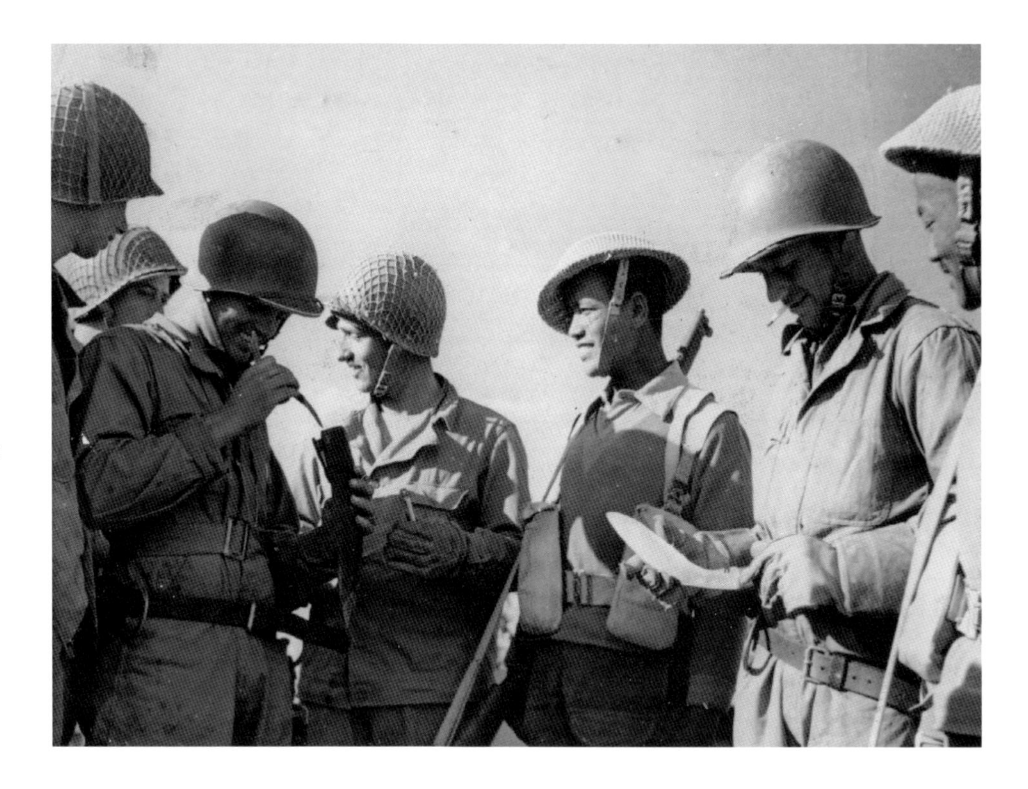

Gurkhas of the 4th Indian Division meet American GIs. The US forces arrived in North Africa as part of Operation Torch which saw some 650 ships put ashore four US and British Divisions on 8 November 1942[30]. The Germans eventually surrendered in North Africa on 12 May 1943 after a three-year campaign. The total German and Italian prisoners amounted to some 250,000

General Sir Claude Auchinleck handed over as commander-in-chief Middle East command in August 1942. In June 1943 he was appointed as the commander-in-chief of the Indian Army, the post he had previously held before deploying to the Middle East. This image shows him visiting recruits in the 8th Gurkhas being trained to operate mortars. The image was taken in Quetta in May 1945

November 1942, had landed in Algeria and Morocco in an operation known as Operation Torch. Montgomery's role, which saw the 1/9th Gurkhas involved in a number of courageous actions, was to prevent the enemy breaking out as Eisenhower's force closed in on them. By 13 May 1943, the last German units had surrendered and fighting in North Africa had ceased.[29]

Gurkhas were also involved in the Allied campaign in Italy. 1/5th Gurkhas, for example, landed on the east coast of Italy as part of 8th Indian Division in October 1943 and took part in a number of notable actions, such as the battles at Atesa and Mozzagrogna, as the Allies advanced north.[31]

The 4th Indian Division arrived in Italy in December 1943 from North Africa.[32] By February 1944, the Division found itself supporting a major Allied offensive to clear the German positions at Cassino. The centrepiece of the German defences was the old monastery of Monte Cassino. It sat on high ground dominating the town of Cassino and its old castle. It also dominated a feature to its immediate front known as Hangman's Hill. Two particular features, known as Snake's Head Ridge and Point 593, were on higher ground than the monastery and were therefore tactically important.

A Gurkha soldier from 3rd Gurkhas shares a quiet moment with an American GI in Italy

Gurkhas were involved in two major assaults on the heavily defended position. The first of these started on 13 February 1944 and involved

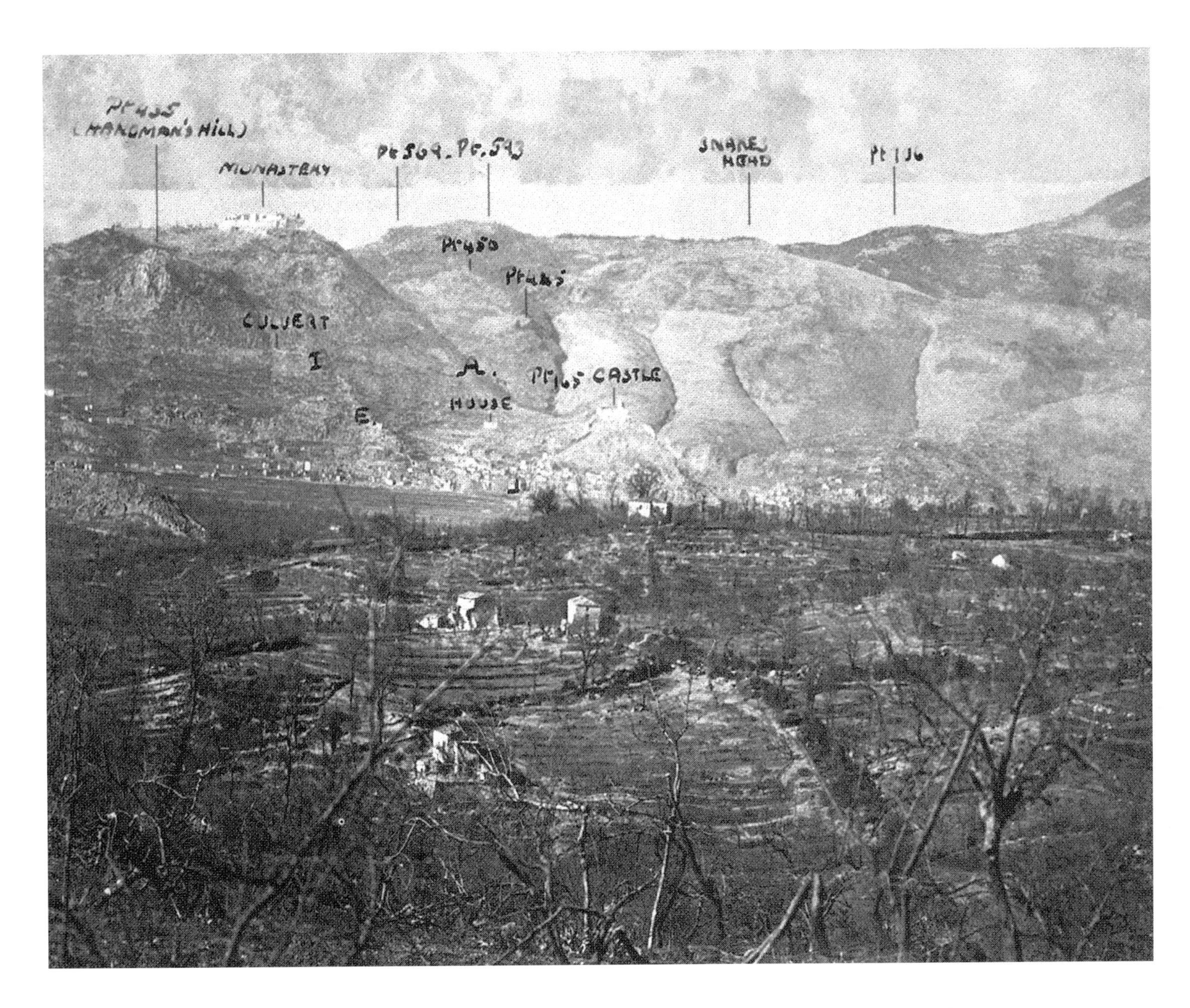

The ground at Cassino with some of the key features highlighted

the 4th Indian Division assaulting forward from Snake's Head to capture Hill 593 before then moving on to take the monastery itself whilst a New Zealand Division assaulted the town from the south.[33] 1/2nd Gurkhas, 1/9th and 2/7th Gurkhas were all in the thick of the action but, despite the Allied forces' heroic efforts, the Germans were just too strong. The Gurkhas sustained considerable casualties with 1/2nd losing over 150 dead and wounded and C and D Companies from 1/9th losing 94 dead and wounded.[34] The second Allied attack on the German defences started on 15 March 1944. The New Zealanders again assaulted the town but this time the 4th Indian Division passed through them to try and take Hangman's Hill and then the Monastery. 1/9th succeeded in capturing Hangman's Hill, a remarkable achievement given the strength of

Hangman's Hill at Cassino with the Monastery of Monte Cassino on the high ground to the rear. On 15 March 1944, 1/9th Gurkhas took part in an Allied assault on the German defences at Cassino, eventually taking Hangman's Hill. They occupied the position for nine days under constant enemy bombardment before being forced to withdraw.

the German defences, but, after nine days of relentless bombardment, they were forced to withdraw.[35] In the event, the German position at Cassino was not taken until May 1944 and then by a Polish Division.[36]

More Gurkha units arrived in Italy during the summer of 1944 as part of 43rd Gurkha Lorried Brigade. Its infantry battalions (the second battalions of the 6th, 8th and 10th Gurkhas) were all soon involved in heavy

A painting by Jason Askew showing 1/9th Gurkhas attacking Hangman's Hill. The battalion succeeded in taking the position. They held it for nine days during which they were subjected to continuous artillery bombardment before having to withdraw

The castle at Cassino with the monastery of Monte Cassino on the high ground to the rear

Soldiers from the 2/4th Gurkhas hitching a lift on one of 6RTR's Churchill tanks in Italy in 1945

This sketch by the artist Harry Sheldon shows Rifleman Sherbahadur Thapa rescuing a wounded comrade having 'dashed forward under accurate small arms and mortar fire.'[39] He was awarded a posthumous Victoria Cross for his actions which took place on the 18–19 September 1944 in San Marino. Sadly, as his citation also states 'while returning the second time he paid the price of his heroism and fell riddled by machine gun bullets fired at point blank range.'[40]

fighting against the Germans in the independent state of San Marino.[37] The 4th Indian Division was also involved in this action with one soldier from 1/9th Gurkhas, Rifleman Sherbahadur Thapa, being awarded a posthumous Victoria Cross for his actions. The digest of his citation reads:

> On the 18th/19th September 1944, a company of the 9th Gurkha Rifles met fierce opposition from a well situated German position. The section commander and Rifleman Sher Bahadur Thapa made a charge and succeeded in silencing the machine gun. After his section commander was wounded, the rifleman, now alone, made his way to the exposed part of a ridge, from here, ignoring the hail of bullets, he managed to silence more machine guns as well as covering a withdrawal and rescuing two wounded men before he was finally killed .[38]

The Germans' last major line of defence in northern Italy was the Gothic Line. This ran along the top of the northern Apennine Mountains and stretched from the west to the east coasts. The Apennines were a formidable natural obstacle in themselves but the Germans had constructed a network of machine gun posts, minefields, trenches and bunkers to further strengthen the defences. All nine Gurkha battalions serving in Italy during the Second World War were involved in fighting to break through the Gothic Line.[41] The battalions fought with courage and determination and there are numerous examples of the bravery of individual Gurkha soldiers. Rifleman Thaman Gurung of 1/5th Gurkhas, for example, was awarded a posthumous Victoria Cross for his actions on 10 and 11 November 1944 at Monte San Bartolo.

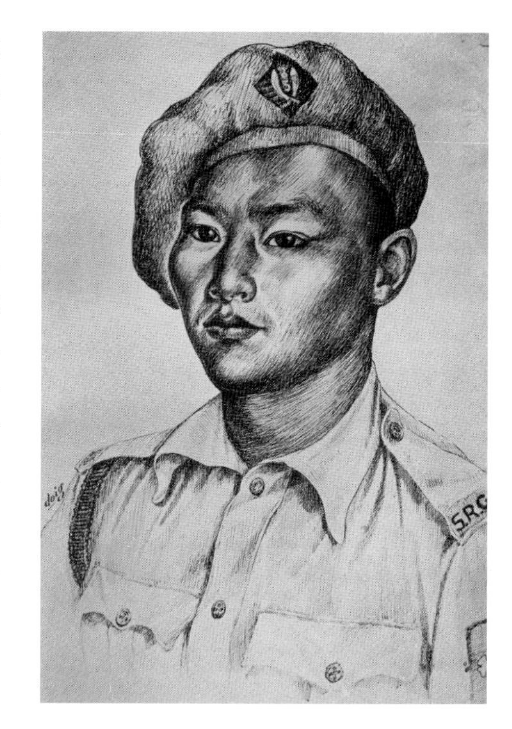

Rifleman Thaman Gurung of 1/5th Gurkhas who was awarded a posthumous Victoria Cross for his actions at Monte San Bartolo in Italy on 10 and 11 November 1944. Rifleman Thaman Gurung was only twenty years old when he was killed.[42] In December 1945, the then Viceroy of India, Field Marshal Lord Wavell, presented Thaman Gurung's mother with the Victoria Cross that had been so heroically won by her son

The fighting continued as the Allied forces broke through the Gothic Line and the Germans withdrew further north. Far from beaten, the Germans had constructed substantial defensive positions along the banks of laterally flowing rivers and canals. These were formidable obstacles and the fighting to seize them was intense and costly, both in lives and equipment. 1/5th Gurkhas, for example, lost 36 killed and 63 wounded in the battle to cross the Senio River in northern Italy.[43]

Fierce fighting also took place in the towns and villages that the Germans temporarily occupied as they continued their fighting withdrawal. In Medicina, for example, 2/6th Gurkhas and 14th/20th King's Hussars

During the advance north through Italy, the Allies encountered spirited resistance from the withdrawing Germans who occupied defensive positions in towns and villages, as well as astride canals and rivers. This image shows Gurkhas fighting from a ruined house during a battle to clear one such defensive position

A painting by Terence Cuneo showing the battle for Medicina in April 1945. 2/6th Gurkhas fought alongside 14th/20th King's Hussars, forging a friendship which continues to this day between The Royal Gurkha Rifles and The King's Royal Hussars.

fought German paratroopers for control of the town.[44] After 24 hours of desperate fighting, the Germans surrendered. The following extract from the Regimental History of the 6th Gurkhas illustrates the nature of the fighting that took place in Medicina on 15 April 1945:

> The scene was in full accord with Wild West cinema technique as the enemy was hunted from house to house. A typical instance was afforded by Subedar Raghu Gurung, who was leading one party down a street supported by a tank. They came under heavy machine-gun fire and a German fired a bazooka at the tank, but missed. The subedar realised that a second shot would probably knock out the tank, so he rushed forward under heavy Spandau fire and killed the German with his kukri and then led his platoon on to capture the strong-point.[45]

His Majesty King George VI being introduced to Jemedar Lalbahadur Rana MC and Naik Kharakbahadur Gurung MM in Italy in 1944

'DIY Bath Unit.' This photograph was probably taken in Italy in 1945 and shows Gurkha ingenuity to maintain personnel cleanliness despite the conditions

Eventually, on 2 May 1945, the German forces in Italy surrendered.

In the early hours of Sunday 7 December 1941, the Japanese attacked the US Pacific Fleet in Pearl Harbor, Hawaii.[46] Three days later and 5000 miles from Hawaii, the British warships HMS Repulse and HMS Prince of Wales, which had been despatched from the British naval base in Singapore to intercept a fleet of Japanese transport vessels in the Gulf of Siam, were attacked and sunk by Japanese aircraft.[47] Japan had entered the war. Ground troops carried by the Japanese fleet were quickly landed in Siam (now Thailand) and the north of Malaya. Conscious of the threat, in September 1941 the British had despatched additional troops to the Malay Peninsula including 2/1st, 2/2nd and 2/9th Gurkhas.[48] But whilst the soldiers of these battalions fought heroically as the Japanese advanced south towards Singapore, there appeared to be little they or any other Allied unit could do to halt the Japanese force. On 15 February 1942, Singapore fell and the remnants of the three Gurkhas battalions were reluctantly taken prisoner.

The fall of Singapore was seen by many as a national disgrace. The Japanese were well trained, highly experienced and appropriately equipped but the British had consistently underestimated them. Moreover, the British defence of Malaya was predicated on a strong naval and air presence. In the event, the Japanese Zero aircraft were more capable than those of the defenders and the sinking of both HMS Repulse and HMS Prince of Wales meant that the Japanese were able

Gurkha soldiers standing next to the graves of two Gurkha riflemen who, taken prisoner by the Japanese, were bayoneted to death by their captors

to land increasing numbers of troops with relative impunity. But whatever the failings of the British defensive plan, the bravery of the soldiers who tried to halt the Japanese invasion remains unquestioned, as does their subsequent conduct in captivity despite the extremes of privation they had to endure. Many were employed building the infamous Burma-Siam 'Death Railway', a Japanese supply route that ran from Bangkok to Rangoon. According to General Sir Walter Walker, whose brother was a prisoner of war and who commanded 4/8th Gurkhas in the later stages of the Burma Campaign, 16,000 of the 61,000 Allied prisoners and 100,000 of 300,000 Asian labourers died building the railway; one man for every sleeper laid.[49]

As well as advancing south into Malaya, the Japanese also advanced west, crossing the Thai border with Burma on 19 January 1942.[50] Their objective was the capital city and key port of Rangoon. To reach Rangoon, the Japanese needed to cross the Sittang River, a major tributary which ran north to south approximately mid way between the capital and the border with Thailand. The only means of crossing the river without using boats was the Sittang Bridge. This had to be held at all costs otherwise the British position in Burma would become untenable. 1/3rd, 1/4th, 1/7th, 2/7th and 2/5th Gurkhas were all involved in defending the bridge and its eastern approaches against numerous Japanese attacks.

This painting by David Rowlands shows men of the 3rd Gurkhas and the Duke of Wellington's Regiment assaulting a Japanese position at the eastern end of the Sittang Bridge. The Japanese were determined to capture the bridge as they advanced towards Rangoon. In the event, it was blown prematurely, leaving some 6000 soldiers from 17th Division trapped between the river and the advancing Japanese

At 0530hrs on 23 February 1942, the bridge was blown, possibly as a result of poor communication[51] or possibly because the divisional commander, Major General J G 'Jackie' Smyth VC, feared that the Japanese would capture it intact if it were not blown before daylight.[52] Whilst blowing the bridge would certainly make life more difficult for the advancing Japanese, it also posed a significant problem for four of the Gurkha battalions (1/3rd, 1/7th, 2/7th and 2/5th Gurkhas) trapped on the eastern side of the river as part of a force of some 6000 men from 17th Division. Approximately half a mile wide, the river was a formidable obstacle, particularly as the steamers, ferries and sampans that might have been used to cross the river had already been destroyed[53] and the fact that the majority of Gurkhas could not swim. Notwithstanding this, the order was given for the defending force to exfiltrate back across the river. There may have been little alternative but it was a costly endeavour with many troops drowning in the fast flowing waters. Of the 2500 troops lost by the 17th Division in its attempts to stop the Japanese advance, about two-thirds were Gurkhas. The impact was significant: 1/3rd Gurkhas were reduced to 221 men (from a total of 750); 2/5th down to 227; and the two battalions of the 7th Gurkhas could only muster about 470 men between them.[54]

There was a chance that, with the support of the Chinese, the Japanese advance might have been contained but, after a series of defeats, the Chinese withdrew their forces back towards their homeland. With the outcome now inevitable, the British took the decision to withdraw their forces north and into British India. Although this sounds relatively straightforward, it was a difficult and close run thing with the Japanese constantly trying to seize the bridges ahead of the British in order to cut off their retreat. The withdrawing force showed tremendous fighting spirit, conducting a series of delaying actions to stall the Japanese advance. 48th Brigade, an all Gurkha organisation, formed the rearguard and had a key role in these actions, buying time to allow the main force to escape. There is little doubt that the heroic efforts of its battle weary and depleted Gurkha battalions contributed significantly to the success of the withdrawal. Eventually, on 21 May 1942, the British force, tired and dispirited after a retreat of over a thousand miles in three and a half months,[55] crossed the border into Assam and British India. It had taken a beating, sustaining a total of 13,463 British, Indian, Gurkha and Burmese casualties compared to Japanese losses of 4597 killed and wounded.[56] Although the Japanese were in hot pursuit, the monsoon, which was just beginning, effectively stopped them from following the withdrawing force across the border. This bought much needed time for the forces that came out of Burma, including the five Gurkha battalions, to be brought back up to strength.

A Gurkha soldier evacuating a wounded comrade in the Arakan. The use of a strap over the head to carry the weight of the load on the back is a familiar technique in the mountains of Nepal. Even today, Nepalese citizens wishing to be recruited into the British Army as Gurkhas have to demonstrate their competence in the technique by carrying a weight of 25kg over a mountainous course of 5km

In July 1942, the British attempted to regain the initiative by launching a counter attack across the border and into the western coastal province of Arakan. Known as the First Arakan Campaign, this achieved little and, having been mauled by the Japanese, the force crossed back into British India in May 1943.[57]

A different approach was needed. It was eventually decided that a specialist force would be formed, known as 77th Brigade, which would be tasked with raiding deep into Burma in order to disrupt the Japanese lines of communication. The force, which comprised some 3000 men, was commanded by Brigadier Orde Wingate and included the recently formed 3/2nd Gurkhas.[58] Known as Chindits after the Chinthe, the mythical half-lion, half-griffin that guards Burmese monasteries and temples, the force was organised into seven columns and began operations on 6 February 1943.[59]

Brigadier (later Major General) Orde Wingate who commanded the Chindits. He was killed when his aircraft crashed during a visit to one of the strongholds established behind enemy lines during the Second Chindit Offensive

Initially, the Chindits made reasonable progress but it was soon apparent that '... the Chindit columns were simply not strong enough to withstand determined Japanese counter-attack, and swiftly turned from hunter to hunted.'[60] The casualties were horrific as the men, now operating in small groups, tried to fight their way back to British India. The Gurkhas, who accounted for 1289 of the 3000 men in the force, fared particularly badly; of the 800 casualties sustained by Wingate's force, 300 were from 3/2nd Gurkhas.[61] But although the first Chindit operations might have achieved little at the tactical level, they were 'beyond price' as propaganda.[62] The Chindits were portrayed as having run rings around the Japanese, helping destroy the myth, which had been reinforced by the failure of the Arakan Campaign, that they could not be beaten in the jungle.[63]

Although the Chindits were operating behind their front lines, the Japanese were keen to push north. From 24 to 27 May 1943, they tried to break through the Chin Hills that lie between British India and Burma.[64]

2/5th Gurkhas found themselves part of the opposing force.[65] Two assaults on a Japanese position known as Basha East Hill had failed

Taken at Imphal in 1943, this picture shows the Viceroy of India inspecting Gurkha survivors of the first Chindit operation

Three soldiers from the 5th Gurkhas who took part in the first of Brigadier Orde Wingate's Chindit expeditions. The soldier in the middle is dressed as a local Burmese villager. Although the first Chindit expedition achieved little tactically, it was presented as a demonstration of how the seemingly invincible Japanese could be beaten at their own game in the jungle

and a third was launched by a force which included a platoon commanded by Havildar Gaje Ghale. Whilst he was preparing for the attack, the havildar was badly wounded by shrapnel from a Japanese grenade. Despite being hit in the arm, chest and leg,[66] he led his platoon in numerous assaults against the Japanese position, inflicting many

A Gurkha soldier preparing a defensive position in the Chin Hills. The photograph was most probably taken in 1943 as the Japanese tried to force their way through the Chin Hills and into British India

enemy casualties and eventually capturing the critical position. He was awarded a Victoria Cross for his actions. One extract from his citation makes notably exciting reading:

> Havildar Gaje Ghale dominated the fight by his outstanding example of dauntless courage and superb leadership. Hurling hand grenades, covered in blood from his own neglected wounds, he led assault after assault encouraging his platoon by shouting the Gurkha's battle-cry.[67]

Havildar Gaje Ghale of 2/5th Gurkhas who was awarded a Victoria Cross for his actions against the Japanese in the Chin Hills in May 1943. He is shown here wearing the rank of Jemedar having been promoted later in 1943. The black cat insignia on his left arm is that of 17th Indian Division.

In December 1943, Lieutenant General Slim, now commanding 14th Army, went on the offensive in what has become known as the Second Arakan Campaign. 5 Indian Division (which included 3/9th Gurkhas), 7 Indian Division (which included 4/1st, 4/5th, 3/6th and 4/8th Gurkhas) and 81 West African Division crossed over the border and, once again, started to advance into the coastal province of Arakan.[68] They encountered stiff resistance from the Japanese who had prepared an impressive line of mutually supporting defences. These exploited the natural obstacle provided by the Mayu mountains and included bunkers linked by tunnel systems. They were formidable positions to take. But the Japanese had no intention of just waiting to be attacked and, in February 1944, they launched a major offensive in the east of Arakan. This was eventually defeated but at considerable cost. 4/1st Gurkhas, for

Soldiers from 10th Gurkhas clearing a trench on a feature known as Scraggy Hill during the fighting around Imphal

Soldiers from the 10th Gurkhas redistributing their equipment after an engagement near Imphal in 1944

example, had been defending two features known as 'Cain' and 'Abel' when they found themselves surrounded as the Japanese advanced. By the time 4/1st eventually left their positions, they had sustained 52 killed and 181 wounded.[69]

The Japanese offensive in the Arakan had been part of a bigger plan to defeat the British and to advance into India. The Japanese had hoped that the British would commit additional units into the Arakan, weakening their defences further east. Their intention was to exploit this, launching the main offensive in the central Assam region in order to seize the towns of Imphal, Kohima and Dimapur.[70]

Although they had taken a beating in the Arakan, the Japanese launched their attack on the Imphal plains in early March 1944.[71] The fighting that followed was fierce and

A Gurkha paratrooper prepares to jump. Two Gurkha parachute battalions were formed during the Second World War: 153 (Gurkha) Parachute Battalion, which later became 2nd (Gurkha) Battalion, Indian Parachute Regiment; and 154 (Gurkha) Parachute Battalion, which later became 3rd (Gurkha) Battalion, Indian Parachute Regiment. 153 was formed at Delhi in October 1941 from volunteers from across all Gurkha Regiments, though the majority came from the 10th Gurkhas. 154 was formed by re-rolling 3/7th Gurkhas in August 1943.[74]

desperate as Slim's forces fought for their lives. The Gurkha battalions were again in the thick of the action. 1/10th, for example, held a critical ridge against sustained Japanese attacks, enabling 17 Indian Division to withdraw to new positions closer to Imphal. 153 (Gurkha) Parachute Battalion, alongside 152 (Indian) Parachute Battalion, fought a desperate battle on 25 and 26 March 1944 against overwhelming Japanese forces at a village called Sangshak.[72] Further north in Kohima, 4/1st Gurkhas played a key role in a three day battle to take a feature known as Jail Hill which dominated the town.[73]

There are numerous stories of heroic actions by Gurkhas during the fighting to halt the Japanese advance and turn the tide in the British favour. One of the most inspirational took place near Imphal on 12 June 1944 at a place called Ningthoukhong. A Japanese force equipped with tanks had succeeded in breaking into a defensive position held by 2/5th Gurkhas. B Company 1/7th Gurkhas were ordered to counter-attack but were caught in the open by machine gun fire from the tanks. Rifleman Ganju Lama, who was carrying an anti-tank gun known as a PIAT (Projector Infantry Anti-Tank) crawled forward and engaged the tanks. He was awarded a Victoria Cross for his actions. This extract from his citation explains why:

> In spite of a broken left wrist and two other wounds, one in his right hand and one in his leg, caused by withering cross fire concentrated upon him, Rifleman Ganju Lama succeeded in bringing his gun into action within thirty yards of the enemy tanks and knocked out first one and then another, the third tank being destroyed by an anti-tank gun. In spite of his serious wounds, he then moved forward and engaged with grenades the tank crews, who now attempted to escape. Not until he had killed or wounded them all, thus enabling his company to push forward, did he allow himself to be taken back to the Regimental Aid Post.[75]

Rifleman Ganju Lama who was awarded a Victoria Cross for his actions against the Japanese near Imphal on 12 June 1944 having already been awarded a Military Medal (MM) for his actions earlier in the Burma Campaign. Born in Sikkim (then an independent kingdom in India) in 1924, Ganju survived the war, eventually retiring from the Indian Army with the Honourary rank of captain. Interestingly, when he retired, he was also appointed as an Honourary ADC to the president of India for life[76]

Two other Victoria Crosses were awarded to Gurkhas during the fighting around Imphal. The first was to Subedar Netrabahadur Thapa of 2/5th Gurkhas for his actions in defence of a key feature known as Mortar Bluff. On 25 June 1944, the feature was attacked by a company strength force from the Japanese 15th Army.[77] Outnumbered and outgunned, Netrabahadur commanded his small detachment of 41 men with courage and determination but eventually the enemy succeeded in taking part of the position. Netrabahadur called for reinforcements and, at 0400hrs the next morning, a section of eight men arrived with ammunition and grenades. Unfortunately, they were all soon wounded. His citation explains what happened next:

> Undismayed, however, Subadar Netrabahadur Thapa retrieved the ammunition and himself with his platoon headquarters took the offensive armed with grenades and khukris. Whilst so doing he received a bullet wound in the mouth followed shortly afterwards by a grenade which killed him outright. His body was found next day, khukri in hand and a dead Japanese with a cleft skull by his side.[78]

The position fell to the Japanese after Netrabahadur's death. Of the 41 Gurkhas who had been defending the position, only six survived.[79] It is therefore perhaps not surprising that Netrabahadur's citation describes Mortar Bluff as '... an epic in the history of the regiment.'[80]

Because of its tactical significance, 2/5th Gurkhas were ordered to recapture Mortar Bluff and the neighbouring position of Water Piquet which had also fallen to the Japanese. C Company began to assault Mortar Bluff but was soon pinned down by enemy machine gun fire.

Jemedar (acting Subedar) Netrabahadur Thapa from 2/5th Gurkhas who was posthumously awarded the Victoria Cross for his actions in defending the feature of Mortar Bluff near Imphal against an overwhelming Japanese force on 25 and 26 June 1944

As Subedar Netrabahadur Thapa was awarded the Victoria Cross posthumously, the medal was presented to his wife, Nainasara Magarni, by the then viceroy of India, Field Marshal Lord Wavell, at a special parade on 23 January 1945. The picture shows Nainasara with the medal

Naik Agansing Rai, commanding one of the lead sections, realised that further delay would cause additional casualties and charged the first of the machine gun posts. He personally killed three of the gun crew before the remainder of the section surged forward and routed the defending enemy.[81] He silenced a second machine gun post in a similar manner before rejoining the rest of his platoon for the assault onto Water Piquet. The citation for the Victoria Cross that he was awarded for his actions explains what happened next:

> In the subsequent advance heavy machine-gun fire and showers of grenades from an isolated bunker position caused further casualties. Once more, with indomitable courage, Naik Agansing Rai, covered by his Bren gunner, advanced alone with a grenade in one hand and his Thompson Sub-Machinegun in the other. Through devastating fire he reached the enemy position and with his grenade and bursts from his Thompson Sub-Machinegun killed all four occupants of the bunker.[82]

Naik Agansing Rai of 2/5th Gurkhas was awarded a Victoria Cross for his bravery on the 26 June 1944 during assaults to capture the key positions of Mortar Bluff and Water Piquet. As his citation notes, his '... magnificent display of initiative, outstanding bravery and gallant leadership, so inspired the rest of the company that, in spite of heavy casualties, the result of this important action was never in doubt.'[83]

In early March 1944, a second Chindit force was inserted behind the Japanese front line. Its aim was to establish strongholds from which the enemy's lines of communication could be disrupted. The intention was to make it difficult for the Japanese to deploy reinforcements in response to a Chinese advance from the north. The plan was ambitious and involved airlanding a large number of troops by glider. Four Gurkha units were involved in the operation: 3/4th; 3/6th; 3/9th; and 4/9th.[84] Operating from their strongholds, the Chindits inflicted considerable damage on the Japanese. The fighting qualities and courage of the Gurkhas were demonstrated on numerous occasions but two actions are worthy of particular mention because they resulted in the award of three Victoria Crosses.

The 5th Gurkhas remained with the Indian Army when India gained Independence in 1947. After this date, medals awarded by the British Government were considered to be 'foreign' awards and therefore lower in precedence than medals awarded by the Indian Government. This explains why the Victoria Cross is in the middle of Honorary Lieutenant Agansing Rai VC's row of medals

The first of these incidents occurred in the town of Mogaung. 77th Brigade, which included 3/6th Gurkhas, had been operating from a stronghold known as 'Broadway' when they were ordered to march 160 miles to take Mogaung in order to support the Chinese advance. Heavily fortified, Mogaung was defended by a force of some 3500 Japanese soldiers. 3/6th Gurkhas were in the vanguard of 77th Brigade's attack. Captain Michael Allmand, then a platoon commander within 3/6th's B Company, was right in the thick of the action. On 11 June 1944, he captured a critical bridge on the outskirts of Mogaung, enabling his brigade to gain a foothold in the town. An extract from his Victoria Cross citation makes remarkable reading:

Captain Michael Allmand of 3/6th Gurkhas who was awarded a posthumous Victoria Cross for his actions at Mogaung during the Second Chindit Operation in June 1944. His citation notes that 'the superb gallantry, outstanding leadership and protracted heroism of this very brave officer were a wonderful example to the whole Battalion and in the highest traditions of his regiment'[87]

> As the platoon came within twenty yards of the bridge, the enemy opened heavy and accurate fire, inflicting severe casualties and forcing the men to seek cover. Captain Allmand, however, with the utmost gallantry charged on by himself, hurling grenades into the enemy gun positions and killing three Japanese himself with his kukri.[85]

Two days later and having taken over command of the company, he led another attack against Japanese snipers, seizing a vital ridge. On 23 June 1944, he led his last attack as his citation makes clear:

> Once again on June 23rd in the final attack on the Railway Bridge at Mogaung, Captain Allmand, although suffering from trench-foot, which made it difficult for him to walk, moved forward alone through deep mud and shell-holes and charged a Japanese machinegun nest, but he was mortally wounded and died shortly afterwards.[86]

As Captain Allmand's company assaulted the bridge, the Japanese opened fire from a location known as 'Red House', pinning the advancing platoons down. One of the lead sections charged towards the house but, in a matter of minutes, Rifleman Tulbahadur Pun found that he was the only unwounded member of the section. Realising the importance of the 'Red House', he grabbed a Bren gun and assaulted it on his own in full view of the defending enemy. As the citation makes clear:

> Despite these overwhelming odds, he reached the Red House and closed with the Japanese occupations. He killed three and put five more to flight and captured two light machineguns and much ammunition. He then gave accurate supporting fire from the bunker to the remainder of his platoon which enabled them to reach their objective. His outstanding courage and superb gallantry in the face of odds which meant almost certain death were inspiring to all ranks and were beyond praise.[88]

Rifleman Tulbahadur Pun who was awarded a Victoria Cross for his actions at Mogaung in June 1944 during the Second Chindit Operation

Mogaung was eventually taken but at considerable cost. As Brigadier Christopher Bullock notes, this might explain why the commander of 77th Brigade, Brigadier Mike Calvert, was irritated to find that the credit for its capture was being claimed by the Chinese force slowly making its way down from the North![89]

The third Victoria Cross was awarded to Major Frank Blaker of 3/9th Gurkhas. On 9 July 1944, Blaker and his company took part in a two company attack on a feature known as Hill 2171.[90] Blaker's C Company

Brigadier Mike Calvert who commanded one of the columns during the second Chindit expedition and Major James Lumley, the father of Joanna Lumley, in the ruins of Mogaung after its capture in June 1944

were caught in the open and engaged by devastating machine gun fire from well prepared Japanese positions. Blaker did not hesitate; his citation states:

> Major Blaker then advanced ahead of his men through very heavy fire and, in spite of being severely wounded in the arm by a grenade, he located the machine guns, which were the pivot of the enemy defence, and single handed charged the position. When hit by a burst of three rounds through the body, he continued to cheer on his men while lying on the ground. His fearless leadership and outstanding courage so inspired his Company that they stormed the hill and captured the objective while the enemy fled in terror into the jungle.[91]

Sadly, Major Blaker subsequently died of his wounds whilst being evacuated from the battlefield.[92]

A Gurkha soldier leading a mule laden with equipment during one of the Chindit operations in Burma

In mid July 1944 the Chindits were recovered back to India. After four months of fighting in the thick of the Burma jungle against a tenacious and ruthless enemy, the break was well deserved. Although media 'gloss' had been applied to portray the First Chindit Operation as successful, there was no need to do this for the second; the actions spoke for themselves.

Having thwarted the Japanese offensives in the Arakan and in Assam, Slim began to move his army south. But the Japanese

Major Frank Gerald Blaker who was awarded a posthumous Victoria Cross for his actions on 9 July 1944 on Hill 2171 near the village of Taungni in Burma. Commissioned into the Highland Light Infantry (City of Glasgow Regiment), Major Blaker served with 3/9th Gurkhas during the Second Chindit Operation. He was awarded a Military Cross for his actions earlier in the Burma Campaign

were far from beaten and contested every mile. As ever, the Gurkhas were in the thick of the action. On 4 March 1945, for example, 3/2nd Gurkhas were involved in a decisive action to secure high ground near the village of Tamandu.[93] One particular feature, known as Snowden East, had been re-captured by the Japanese and dominated the route along which 82nd West African Division was attempting to evacuate its casualties.[94] B Company of 3/2nd Gurkhas was ordered to re-take the feature. As the company advanced, it was engaged by a Japanese sniper hiding in a tree. Unable to fire from the prone position, Rifleman Bhanbhagta Gurung stood up in full view of the enemy and 'calmly killed the enemy sniper with his rifle, thus saving his section from further casualties.'[95] But Bhanbhagta didn't stop there. As the Japanese engaged the advancing company from well prepared foxholes, Bhanbhagta dashed forward alone and attacked the first foxhole, throwing two grenades and killing its two occupants. He charged forward to the next foxhole, killing its occupants with his bayonet. He took a further two foxholes before changing tack and attacking a machinegun post which, located in a bunker, would have been able to engage both his own platoon as well as another one advancing from the west. Out of high explosive hand grenades, he lobbed two smoke grenades through the slit and, as the Japanese emerged, he killed them with his kukri. He then crawled inside the bunker to despatch the remaining enemy soldier. It was a remarkable achievement, fully deserving of the Victoria Cross which he was awarded. As his citation notes:

> Rifleman Bhanbhagta Gurung showed outstanding bravery and a complete disregard for his own safety. His courageous clearing of five enemy positions single-handed was in itself decisive in capturing the objective and his inspiring example to the rest of the Company contributed to the speedy consolidation of this success.[96]

Slim continued to advance south. Knowing that the Japanese intended to mount a spirited defence around the town of Mandalay, which lay on a bend of the mighty Irrawaddy River, Slim decided to cross the river to

Rifleman Bhanbhagta Gurung who was awarded the Victoria Cross for single-handedly clearing five heavily defended enemy positions on the Snowden East feature near the village of Tamandu in Burma on 5 March 1945. Rifleman Bhanbhagta left the army immediately after the war, returning to his village in Western Nepal. His Company Commander (later Colonel D F Neill OBE MC) described him as '... a smiling, hard swearing, gallant and indomitable peasant soldier, who, in a battalion of very brave men, was one of the bravest.'[97]

Gurkhas crossing the mighty Irrawaddy River in central Burma as the 14th Army, under the command of Lieutenant General Sir William Slim, advanced south towards Rangoon

the west and north of the city, cutting the Japanese lines of communication and enabling him to seize vital airfields further south.[98]

Slim's forces crossed the Irrawaddy in January and February 1945. The Japanese put up a spirited defence, particularly in Mandalay, but eventually withdrew, evacuating the city to avoid being cut-off. As the Japanese retreated, Slim inserted his own forces ahead of them to block their retreat. The idea was that this would buy sufficient time for his main forces to catch up with and destroy the withdrawing enemy.[99] One such delaying action involved 4/8th Gurkhas and occurred at a place called Taungdaw. Rifleman Lachhiman Gurung and his section were manning an isolated post located about 100 metres forward of the main company defences.[100] At 0120hrs on 12 May 1945, some 200 Japanese attacked the company position. Rifleman Lachhiman Gurung's section bore the brunt of the attack. The Japanese hurled numerous grenades at Lachhiman's trench in an attempt to try and clear a route onto the main company position. Lachhiman picked two of the grenades up and hurled them back. He tried to do this with a third grenade but it '... exploded in his hand, blowing off his fingers, shattering his right arm and severely wounding him in the face, body and right leg.'[101] The enemy then attacked in force but Rifleman Lachhiman Gurung '... regardless of his wounds, fired and loaded his rifle with his left hand, maintaining a continuous and steady rate of fire.'[102] Rifleman Lachhiman's trench

Riflman Lachhiman Gurung of 4/8th Gurkhas who was awarded a Victoria Cross for his actions in defending an isolated position against repeated Japanese attacks on 13 May 1945. Rifleman Lachhiman was just under five feet tall. Had recruiting standards not been relaxed because of the war, he would have been too short to enlist into the Gurkhas![103]

Rifleman Asman Gurung of the 6th Gurkhas crossing the Irrawaddy

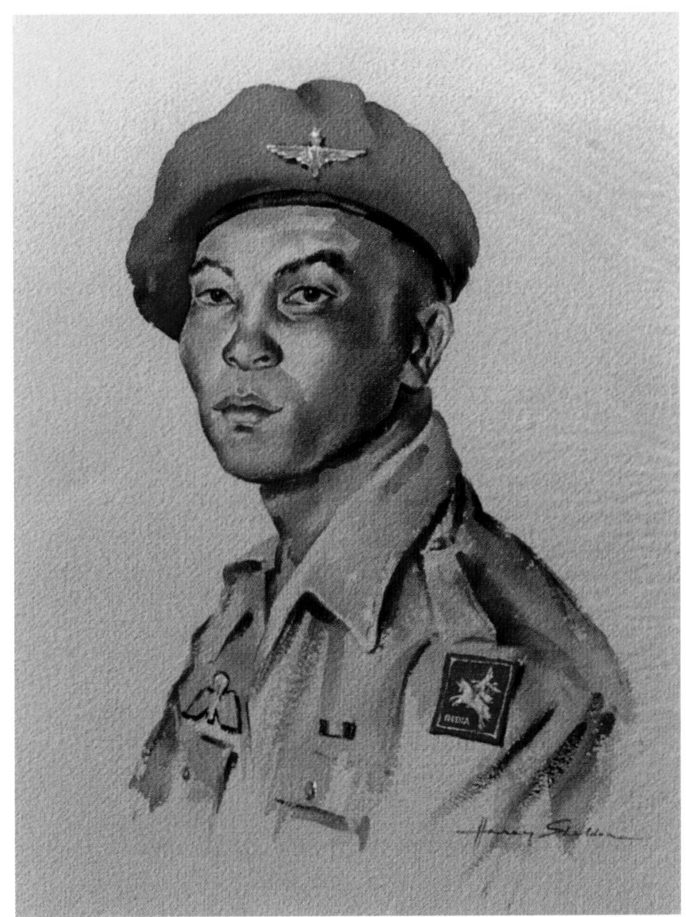

A portrait of Company Havildar Major (later Major (Gurkha Commissioned Officer)) Bhimbahadur Thapa who was a PT instructor at the Indian Airborne Forces Training Unit in Chaklala

was the key to the whole company position. Had the Japanese taken it, the company position would have been quickly lost. It is therefore no surprise that he was awarded the Victoria Cross.

On 3 May 1945, the British retook Rangoon in Operation Dracula, an air, land and sea assault against the city and its defences. It was highly successful, prompting Churchill to coin the term 'triphibious' to describe the operation.[104] A few days earlier on 1 May, the 50th Indian Parachute Brigade, which included a composite Gurkha parachute battalion, dropped on Elephant Point (where the Irrawaddy discharges into the sea) to destroy the Japanese coastal artillery that dominated the sea approach to Rangoon.[105,106]

Nazi Germany surrendered on 8 May 1945 but, although the British had retaken Rangoon five days earlier, the fighting in Burma continued as some 20,000 Japanese soldiers tried to escape east across the

Soldiers from the 10th Gurkhas escorting Japanese prisoners of war

border with Thailand.[107] To make good their escape, they had to cross the mighty Sittang River which, at the time, was in full flood and about as wide as 'the Thames at London Bridge.'[108] The author John Masters, who commanded 111 Brigade during the second Chindit Operation and who served as the senior staff officer in 19 Indian Division during the latter stages of the Burma Campaign, explains what happened as the Japanese left the relative safety of the Yomas mountains and started to move east:

> They came on, and out of the sheltering Yomas. The machine guns got them, the Brens and rifles got them, the tanks got them, the guns got them. They drowned by hundreds in the Sittang, and their corpses floated in the fields and among the reeds. In July 1945 we of the 14th Army killed and captured 11,500 Japanese for a loss of 96 killed.[109]

With the Japanese now a spent force in Burma, Slim's Army expected to be redeployed to Malaya. However, on 6 August 1945, a Boeing B-29 Superfortress belonging to the US Air Force dropped the atomic bomb 'Little Boy' on Hiroshima. Three days later on 9 August 1945, the US dropped a second bomb, this time on Nagasaki. The devastation caused by these two bombs was unprecedented and, on 15 August 1945, Japan announced its surrender. The actual Instrument of Surrender was signed on board the USS Missouri in Tokyo Bay on 2 September 1945, bringing the Second World War to an end.

The Japanese surrender in Rangoon in May 1945. The picture shows Lieutenant General Namata surrendering to Major General W S Symes, the General Officer Commanding South Burma

As this chapter illustrates, Gurkhas served with distinction throughout the Second World War in numerous theatres. But their role in the Burma Campaign as part of Slim's 'Forgotten 14th Army' deserves particular mention. 35,000 Gurkhas served in Burma[110] and officers and soldiers from Gurkha regiments were awarded nine Victoria Crosses during the campaign. These remarkable statistics illustrate the important contribution that Gurkhas made to the success of this most demanding campaign.

Taken after the end of the Second World War, this picture shows three winners of the Military Cross from the 3rd Gurkhas

Bands within the Brigade of Gurkhas

The Band of the 8th Gurkhas in 1886

The Brigade of Gurkhas has a long tradition of military music. The first military band was formed by the Sirmoor Rifle Regiment (later the 2nd Gurkhas) in 1859, although it was another 20 years before regimental bands were formally recognised by the Indian Army. At the outbreak of the First World War, the majority of Gurkha Infantry Regiments had already established their own pipe and military bands.

When India achieved independence in August 1947, the 2nd, 6th, 7th and 10th Gurkhas all transferred to the British Army whilst the 1st, 3rd, 4th, 5th, 8th and 9th Gurkhas remained as part of the Indian Army. As explained in Chapter 7, soldiers in those Regiments destined to join the British Army were given the option of staying with their regiment or of transferring to a regiment which would remain in the Indian Army. Perhaps surprisingly, a large number of Gurkhas opted to remain in the Indian Army. Several reasons have been given for this. One is that many Gurkhas did not wish to leave the familiarity of the Indian Army for what were, at the time, uncertain conditions of service in the British Army. Another is that joining the British Army meant serving away from the Indian subcontinent and therefore further away from Nepal. Whatever the reason, very few musicians opted to transfer to the British Army with their regiments. These regiments were therefore given the choice of reforming either a military band or a pipe band. The

The Band of the 1st Gurkhas circa 1913. At the outbreak of the First World War, the majority of Gurkha regiments already had their own military bands

A Gurkha regimental band in Abbotabad, 1927

The Band of The Brigade of Gurkhas taking part in the Edinburgh Military Tattoo in 1997

6th, 7th and 10th Gurkhas chose to re-form their pipe bands. The 2nd Gurkhas, who had always had bugles rather than pipers because of their close affiliation with the 60th Rifles, opted to re-raise a military band.

In 1955, the Brigade of Gurkhas was permitted to raise a Staff Band in addition to the Regimental Band of the 2nd Gurkhas.[1] However, at the end of the Borneo Confrontation, the Brigade of Gurkhas was reduced in strength from 15,000 to 7000. One consequence of this was the amalgamation in Singapore of The Staff Band, The Brigade of Gurkhas with the Regimental Band of the 2nd Gurkhas. The new composite band, which had the status of a Minor Staff Band, was badged to the 2nd Gurkhas and became the Band of The Brigade of Gurkhas (2nd King Edward's VII's Own Gurkha Rifles). Of note, the band deployed on operations during the First Gulf War in 1991, forming the Gurkha Ambulance Group with 28 (Ambulance) Squadron, The Gurkha Transport Regiment (GTR).

On 1 July 1994, the 2nd, 6th, 7th and 10th Gurkhas amalgamated to form The Royal Gurkha Rifles. The band also rebadged to the new regiment and changed its name to the Band of The Brigade of Gurkhas.[2]

The son of a Gurkha soldier tries his hand at the trumpet during a concert by the Band of The Brigade of Gurkhas. The concert took place at the Hornbill School in Brunei in 2010

Pipers from The Royal Gurkha Rifles often perform with the band, creating a spectacular display

Previous page Members of the Band of The Brigade of Gurkhas at Dover Castle in 2012

Above Drums of the Band of The Brigade of Gurkhas

The Band of The Brigade of Gurkhas giving a concert in Cyprus. The band works hard to make its concerts interesting by including traditional Nepali dances, including the kukri dance, in its programme

Bandsmen within the Band of The Brigade of Gurkhas are accomplished musicians and attend the same courses as other bandsmen within the British Army at the Royal Military School of Music at Kneller Hall

The musician Jules Holland with members of the band during a concert in support of the Army Benevolent Fund in 2009

The Band of The Brigade of Gurkhas taking part in a military tattoo in Virginia, USA. The band plays an important role in raising the profile of the Brigade of Gurkhas, both nationally and internationally

The Band of The Brigade of Gurkhas is one of the 24 military bands that now exist in the British Army. It is led by a Director of Music from the Corps of Army Music and comprises of 44 Gurkha musicians. Before joining the band, Gurkha musicians must complete the same basic training as every other Gurkha at the Infantry Training Centre in Catterick. Once they join the band, they, like other musicians in the British Army, attend courses at the Royal Military School of Music at Kneller Hall. This ensures that they continue

Musicians within the band are able to play a selection of traditional Nepalese instruments as well as their usual instruments

to develop professionally as musicians. The band has a busy schedule of concerts and performances both nationally and internationally and is acknowledged as one of the Army's most exciting bands.

Although it is one of the Army's premier concert bands, the Band of The Brigade of Gurkhas continues to have an operational role in support of the Medical Services. Its specific task is chemical decontamination, a skill it maintains through regular training.

Members of the Band of The Brigade of Gurkhas in conversation with The Colonel-in-Chief of The Royal Gurkha Rifles, His Royal Highness The Prince of Wales, during a medal parade for 1RGR in 2011

CHAPTER 7

The Post-War Years and Malaya

1946–1960

Some 138,000 Gurkhas served the British Crown during the Second World War,[1] organised into 45 Gurkha infantry battalions.[2] Of these: 7539 were either killed or died of wounds or disease; 14,082 were wounded but survived; and a further 1441 were posted as missing, presumed dead.[3] By any measure, these figures are remarkable and provide an indication of the outstanding contribution that Gurkhas made to Britain's war effort, receiving 2760 awards for bravery or distinguished service, including 12 Victoria Crosses, 77 Distinguished Service Orders (DSO) and 333 Military Crosses (MC).[4] But whilst many regiments were able to enjoy a degree of peace and tranquillity after the war, this was not the case for the Gurkhas. With some units already deployed in the Far East, they soon found themselves yet again in the thick of

Japanese soldiers crossing a river with arms, ammunition and equipment for surrender to 1/10th Gurkhas in 1945. In the months immediately after the end of the Second World War, Japanese soldiers and Gurkhas fought alongside each other against nationalist insurgents in Indo-China

the action, fighting insurgents in Malaya, Indo-China and Indonesia. They also had to overcome an entirely unexpected challenge as India achieved independence and six of the ten Gurkha Regiments found themselves remaining in the Indian Army whilst only four transferred to the British Army.

Perhaps surprisingly given they had spent the last few years fighting them, the final months of 1945 saw Gurkhas working with the Japanese as, together, they fought to maintain order in Indo-China until French forces could be deployed to take control of their erstwhile colony. The enemy was the nationalist Annamite movement, later to be more widely known as the Viet Minh. Skilled jungle operators, there were frequent clashes between the Viet Minh and British and Japanese forces. 1/1st, 3/1st, 4/2nd, 3/8th and 4/10th Gurkhas were all involved in the conflict which, although primarily low level, occasionally required deliberate attacks by brigade-sized forces.[5] The Gurkhas deployed to Indo-China in September 1945 and remained there until they were relieved by French forces in January 1946.[6] Although short in duration, the Gurkhas' contribution to stability in Indo-China was significant. It was also costly; the two battalions of the 1st Gurkhas, for example, lost 14 killed and 28 wounded during the three month conflict.[7]

The Gurkhas also found themselves fighting in the jungles of Indonesia against the nationalist forces of Dr Ahmed Sukarno. Six Gurkha

A Gurkha soldier standing guard over Indonesian prisoners in 1945. Gurkhas operated against nationalist terrorists opposed to the re-imposition of Dutch colonial rule for about twelve months until relieved by Dutch troops

battalions were involved: 3/3rd, 3/5th, 1/8th, 4/8th, 3/9th and 3/10th.[8] Their primary task was to protect Dutch and Eurasian internees in the many prisoner of war camps established by the Japanese during their occupation. Notwithstanding the nature of the task, the fighting was often intense, involving tanks, artillery, naval gunfire and air support.[9] Eventually, after nearly a year of constant combat, the Gurkhas were re-deployed back to India as Dutch troops arrived to take over the task.[10]

On 15 August 1947, India eventually achieved independence. A momentous occasion for the subcontinent, it also brought with it huge changes for the Gurkhas. About a week before the announcement was made, the British Government had determined which of its Gurkha regiments would transfer to the British Army and which would remain in the Indian Army.[11] The eventual decision, which was largely based on the proximity of each regiment to the Gurkhas' new home in Malaya, saw 2nd, 6th, 7th and 10th join the British Army whilst 1st, 3rd, 4th, 5th, 8th and 9th Gurkhas remained in the Indian Army.[12] Soldiers in those regiments selected to join the British Army were able to 'opt' to remain in their regiment or to transfer to a regiment which would stay in the Indian Army.[13] If neither of these options appealed, they could also opt to be discharged from military service and return to Nepal, although this option subsequently appears to have been rescinded.[14] In the event, a surprisingly large number of soldiers serving in regiments due to join the British Army opted to transfer to Indian Army units. There were a number of possible reasons for this unexpected outcome. One reason was that many Gurkhas did not wish to leave the Indian Army for what, at the time, were uncertain conditions of service in the wider British Army.[15] Another reason was that India was offering full commissions to its Gurkha officers rather than the lesser 'Viceroy's' commissions available in the British Army. As a result, many senior ranks opted for the Indian Army and persuaded their men to do likewise. Whatever the reason, the inevitable consequence was that the units transferring to the British Army were worryingly short of trained manpower. The average strength of each battalion on transfer appears to have been around 300 men, about a third of the full compliment.[16] John Cross notes that 2/7th Gurkhas came across with only about 100 men[17] whilst Brigadier Christopher Bullock notes that of the 4288 badged 6th Gurkhas, only 452 opted for service in the British Army.[18] In the latter case, the numbers were partially made up by re-enlisting 390 wartime soldiers who had recently been discharged from the 6th Gurkhas.

Soldiers from the 7th Gurkhas firing 25-pounders. In 1948, two battalions of the 7th Gurkhas were hastily converted into artillery regiments as part of a drive to form an all-Gurkha division in the Gurkhas' new home location of Malaya. However, after 13 months the experiment was abandoned and both battalions returned to the infantry role

By April 1948, the majority of the British Army's Gurkha units had been redeployed to their new base locations in and around Malaya.[19] It was an interesting time, particularly for 1/7th and 2/7th Gurkhas. As part of a drive to form an all-Gurkha Division in Malaya, these two units were hastily turned into gunner regiments. Officers from the Royal Artillery were posted in and the units were issued with 25 pounder guns.[20] However, the experiment was not a great success and after 13 months of being known as 101 and 102 Field Regiments Royal Artillery (from June 1948 to July 1949)[21] both battalions were converted back into Infantry units.[22]

Gurkha signallers laying line in the jungles of Malaya in 1949. The Gurkha signals were formed in 1948 as part of the drive to form 17 Gurkha Division. The regiment's title changed to its current name of the Queen's Gurkha Signals in 1977

Having arrived in their new locations, the Gurkha regiments were looking forward to a period of relative quiet in which to reorganise after the turbulence of independence. Units of Gurkha engineers (later to become The Queen's Gurkha Engineers) and Gurkha signallers (later to become the Queen's Gurkha Signals) were formed, as was a unit of Gurkha Royal Military Police.[23] Meanwhile, the rifle regiments turned their attention to training the many new recruits who were arriving from Nepal to fill their depleted ranks. But the peace and tranquillity did not last long. On 17 June 1948, the British high commissioner

Despatch riders of the Gurkha Signals circa 1949 in Malaya

declared a state of emergency in response to the murder of three British rubber planters.[24] This time the enemy was the Malayan Communist Party (MCP) and its action arm, the Malayan Races Liberation Army (MRLA).[25] Comprised of some 5000 armed men, the MRLA was organised into ten district regiments and was supported by about 250,000 local Chinese (known as the Min Yuen).[26]

The leader of the MRLA was Chin Peng. Supported by the British, he had fought against the Japanese during the war, receiving an MBE for his efforts. The communist terrorists (CT) of the MRLA adopted guerrilla tactics, ambushing 'soft skinned' vehicles and attacking lightly guarded plantations and tin mines owned by Europeans.[27] They were excellent jungle fighters and 'The Emergency' as it was known became a 'deadly game of hide and seek' with the security forces as the hunters and the CT as the hunted.[28] The Gurkhas were deployed into the jungle for weeks at a time in an attempt to separate the CT from their supporters. But the CT were no fools and they managed to achieve some notable successes against the British who, to some extent, were still coming to terms with the nature of the conflict. In January 1949, for example, the CT ambushed a platoon from A Company 1/6th Gurkhas near the Thai border.[29] Major Ronald Barnes, Captain (King's Gurkha Officer (KGO)) Tulprasad Pun and nine other Gurkhas were killed.[30]

Gurkha engineers transporting a light gun across a river in Malaya during the Emergency. The Gurkha engineers were formed in 1948 as one of the arms and services within 17 Division, which was originally intended to be an all-Gurkha division. The regiment changed its title to its current name of The Queen's Gurkha Engineers in 1977

Gurkha engineers on patrol in a rubber plantation in Malaya during The Emergency

On 1 March 1949, the Malay Government introduced a policy of national registration.[31] Although a significant administrative burden, this provided the government with an understanding of how the population was distributed throughout the country. On 28 May 1949, the government followed this up by passing legislation which enabled it to forcibly resettle the Chinese population of Malaya in so-called 'new villages' which could be defended against the CT.[32] In doing this, the intention was to isolate the CT from their traditional support base, forcing them to become increasingly self-reliant.

As they gained in experience, the Gurkhas began to score some notable victories against the CT. One such success occurred in January 1950 in Jahore, the southern-most state of the Malay Penninsula. Over the previous 12 months, 1/2nd Gurkhas had been largely frustrated in their attempts to engage the local MRLA unit. Known as 7 Company, they were arguably the most ruthless and militant CT grouping.[33] On 5 January 1950, the terrorists attacked and derailed a train, killing a number of the occupants. The immediate follow-up operation failed to achieve results but information was subsequently received suggesting where the terrorists might be based. On 22 January 1950, Major Peter Richardson and two platoons of B Company, 1/2nd Gurkhas were despatched into the jungle to find the terrorists.[34] Arriving near where they believed the terrorists to be hiding, they spread out and, as dawn broke, started to advance through the swamp and paddy fields ahead of them. Suddenly, a terrorist appeared, running towards a squatter's hut about 50 yards in front of the advancing forces. He was immediately engaged by the Gurkhas but, alerted by the sound of gun shots, a number of other CT appeared and returned the Gurkhas' fire.[35] What followed has been described as the 'largest contact in the whole Emergency' as the Gurkhas outflanked and eventually defeated the terrorists.[36] When

A Gurkha from 7th Gurkhas watches as local Malays in the northern state of Perak demonstrate their expertise with the blowpipe

the battle was over, the Gurkhas found the bodies of 12 terrorists, although it would later transpire that 35 terrorists had either been killed or mortally wounded, including the notorious commander of 7 Company MRLA.[37] Major Richardson was awarded the Distinguished Service Order (DSO) for his leadership during the operation and his sergeant major, Warrant Officer Second Class Bhimbahadur Pun, the Distinguished Conduct Medal (DCM).[38] This incident was typical of the operations carried out by the Gurkhas against the CT throughout The Emergency.

Coordination between the various Government authorities involved in The Emergency started to improve with the formation of State and District War Executive Committees (SWEC and DWEC).[39] With appropriate representation from the civil administration, the military and the police, these committees proved invaluable in ensuring that the efforts of the various state actors involved in The Emergency were mutually supporting. Although the authorities adapted their approach as they learnt from their experiences, the terrorists continued to achieve some successes. On 17 August 1951, for example, two platoons of A Company, 2/7th Gurkhas were patrolling near to their base location when they were ambushed by some 80 CT.[40] The Gurkha force was commanded by Lieutenant (KGO) Dhanbahadur Gurung MC IDSM. A highly experienced and capable soldier, Dhanbahadur has been described as '... a man of unusual forcefulness, tactical talent, and courage, who was respected and feared by his men.'[41] The Gurkhas were pinned down by the weight of fire coming from the terrorists and, in particular, from their five machine guns. Dhanbahadur personally led three charges to try and take the CT positions but without success. Throughout the contact, he continued to encourage his men by walking up and down their line despite being in full view of the enemy.[42] Eventually, just as the CT were about to launch an attack, a small force commanded by Major E Gopsill DSO MC arrived to support Dhanbahadur. They were just in time; dangerously low on ammunition and realising that they were about to be over-run, Dhanbahadur had already ordered his men to draw their kukris![43] Dhanbahadur received a bar to his MC for his bravery and two other Gurkhas were awarded the Military Medal (MM) for their actions.[44]

In October 1951, the British High Commissioner in Malaya, Sir Henry Gurney, was ambushed and killed.[45] Later that same year, the architect of the plan to defeat the CT, General Sir Harold Briggs, retired, dying shortly afterwards. One might suppose that these two events would be

General Sir Gerald Templar inspecting a detachment of 6th Gurkhas. In October 1951, Winston Churchill, the British prime minister, appointed Templar as both the British high commissioner in Malaya and director of operations, unifying the civil and military command chains

catastrophic for the campaign in Malaya but, in the event, they shocked Whitehall into appointing a single man as both the high commissioner and director of operations.[46] The man selected for the job was General Sir Gerald Templar, a highly experienced and capable military commander who, during the Second World War, had commanded divisions in both North Africa and Italy. As Charles Allen notes, in being both high commissioner and director of operations, Templar would have 'powers that no British soldier had ever had since Cromwell.'[47]

Templar's arrival had immediate impact.[48] An energetic and determined officer, he realised the importance of undermining the MRLA's cause by getting the UK Government to agree to Malaya's independence. He also realised the importance of gaining the support of the indigenous population, widening the membership of the regional War Executive Committees to include prominent civilians and coining the phrase 'winning the hearts and minds of the people'.[49] Slowly, the security forces started to gain the upper hand. But it was hard work. As John Cross notes, '... for every million hours of security force endeavour in the Federation of Malaya the enemy was in the sight of a soldier's weapon for 20 seconds.'[50] Operations continued with notable successes being achieved by the Gurkhas as the CT were driven ever deeper into the jungle. In 1955, the country held its first democratic elections. These were won by the Triple Alliance Party headed up by Tunku Abdul Rahman. The new administration attempted to reconcile the CT, with talks being held between Tunku Abdul Rahman and Chin Peng (the leader of the Malay Communist Party), but these failed to achieve a satisfactory conclusion.[51]

Operations against the CT continued, often from security force 'forts' located deep in the jungle.[52] Not only did these forts restrict the CT's freedom of manoeuvre but they also reassured the local aborigines, something that paid significant dividends as the locals started to pass on information about CT activities.[53] On 2 January 1956, Major (Gurkha Commissioned Officer (GCO)) Harkasing Rai MC IDSM MM set out with C Company 1/6th Gurkhas to investigate one such snippet of information regarding the movements of about 50 CT.[54] The Gurkhas tracked the CT for just over a week before eventually making contact. Although the majority of the terrorists escaped across a river, the Gurkhas managed to kill three of them.[55]

The Gurkhas became expert at patiently tracking their quarry through the dense jungle. The CT, aware of this, worked hard to make the Gurkhas' task as difficult as possible. On one occasion in March 1956, C Company, still commanded by Major (GCO) Harkasing Rai, again

The mortar platoon of 1/10th Gurkhas supporting an attack by A Company of the same battalion against communist terrorists in Malaya

A jungle river patrol being carried out by 10th Gurkhas in Malaya during the Emergency. Then, as now, rivers provided relatively easy access to remote jungle areas. Today's Gurkhas maintain this essential riverine capability in Brunei

Major (GCO) Harkasing Rai the Officer Commanding C Company, 1/6th Gurkhas wearing his MC and bar. The bar was awarded for his actions against communist terrorists in Malaya in 1956. The painting in the background is of Field Marshal Viscount Slim who served with 1/6th as adjutant and later commanded 2/7th Gurkhas before going on to establish a reputation as a brilliant wartime general

found themselves tracking CT over the most inhospitable terrain. Following the line of a steep ridge, they eventually located a small CT camp with about 15 terrorists.[56] After spending an uncomfortable night lying within metres of the terrorist base, C Company tried to surround the camp but were spotted. A fierce exchange took place as the terrorists split into small groups and fled.[57] For his actions against the CT, Major (GCO) Harkasing Rai was awarded a bar to his MC.

On 31 August 1957, Malaya became an independent country within the Commonwealth.[58] The fighting continued but, by the end of 1958, the CT had been all but defeated. The 12 year campaign against Chin Peng's communist guerrillas formally came to an end on 1 August 1960.[59] That the Gurkhas had made a remarkable contribution to the successful outcome of the counter-insurgency campaign is not disputed. Not only did Gurkha units account for '... over a quarter of all eliminations during The Emergency' but they conducted themselves in a manner which earned the respect and gratitude of the Malayan people.[60] They also demonstrated to the wider British Army that, when it came to jungle operations, they had no equals. Interestingly, the Gurkhas' natural empathy with the indigenous population and keen eye for the evidence of human transit through an area of jungle (known as 'sign' in tracking parlance) have proven to be just as important on contemporary operations in Afghanistan as they were in Malaya.

Detachments from both Battalions of the 6th Gurkhas take part in a Victory Parade in Kuala Lumpur on 1 August 1960 to mark the end of the Malayan Emergency.

Gurkha Paratroopers

A copy of a watercolour sketch by the artist Harry Sheldon of a Gurkha paratrooper. The padded helmet, which was made of hessian, was worn to prevent impact damage to the head on landing. They had to be constructed locally as there were no proper helmets available for the 14th Army which was always at the bottom of the equipment priority list!

The first Gurkha parachute unit was formed in Delhi in October 1941 by Lieutenant Colonel (later Major General) F J Loftus-Tottenham. Titled 153 (Gurkha) Parachute Battalion, it was made up of volunteers from across all the Gurkha regiments, although the majority came from the 10th Gurkhas.[1] This was probably because Loftus-Tottenham had been commanding 1/10th before being selected to raise the Gurkha Parachute Battalion.[2]

The battalion's initial parachute training took place in India. In commenting on how the Gurkhas took to this, Loftus-Tottenham makes an interesting observation:

> The nastiest part about parachuting is landing, and in this, with his compact light body and strong hill legs, he has a distinct advantage over most other races. For this reason I would say that, along with the Japanese who is similarly built, he is probably the best natural parachutist in the world.[3]

The first operation conducted by Gurkha paratroopers from 153 (Gurkha) Parachute Battalion took place in Burma in July 1942 and was known as Operation

Major (later Lieutenant Colonel) J O M 'Jimmy' Roberts who commanded the first Gurkha parachute operation. Known as Operation 'Puddle', this took place in Burma and involved a small group of Gurkhas and British soldiers parachuting in behind enemy lines to gather information about Japanese dispositions. Major Roberts was awarded a Military Cross for his actions. An accomplished mountaineer, he achieved a number of notable first ascents in the Himalayas. He made several attempts to reach the summit of Everest and, in 1960, was the expedition leader of the first successful ascent of Annapurna II.[4]

Rifleman Dilbahadur Thapa. Originally from 3/1st Gurkhas, Rifleman Dilbahadur joined 153 (Gurkha) Parachute Battalion and served in Burma, acting as Major Jimmy Roberts' orderly. He was one of the battalion's many killed in action at the Battle of Sangshak in March 1944

'Puddle'.[5] Commanded by Major J O M 'Jimmy' Roberts, this involved eight Gurkha and three British paratroopers jumping in behind enemy lines in order to '... carry out reconnaissance and make certain contacts behind or on the flank of the Japanese lines in Burma.'[6] After six weeks, the group eventually arrived at a location known as Fort Hertz. The second parachute operation conducted by 153 (Gurkha) Parachute Battalion took place in August 1942 and was known as Operation 'Firepump'. This consisted of dropping another small group of Gurkha and British paratroopers, including some military engineers, into Fort Hertz.[7] The plan was that this group would meet up with Roberts' team and then prepare a landing strip so that a British Infantry company could be flown in to garrison the Fort.[8] Both operations were successful and the paratroopers were recovered by the same aircraft that eventually delivered the infantry company. In September 1942, Major Roberts was awarded the Military Cross for his leadership of Operation 'Puddle'.[9]

153 (Gurkha) Parachute Battalion on parade in Delhi 1942

The need for an additional Gurkha parachute battalion was soon identified and it was decided to re-form 3/7th Gurkhas in the parachute role in May 1942. 3/7th had been amalgamated with 1/7th in February 1942 following losses sustained by the latter in the fighting at Sittang Bridge (see Chapter 6). On 4 August 1943, the newly re-formed 3/7th Gurkhas changed its name to 154 (Gurkha) Parachute Battalion.[10] There is an amusing anecdote about the selection of volunteers for this second unit which is worth repeating:

> At once all the officers set about explaining what parachuting involved – not easy to do since they themselves had only sketchy ideas on the subject. However they managed to get hold of a number of films about parachutists and these were shown to the men. There was no time to run the films through beforehand. The first film, shown to a packed audience, began with the words: 'now if you do your job well there is no reason why ninety-five percent of these men should ever reach the ground alive.' It was a film designed to teach ground troops how to deal with a parachute attack! It was a great success. All that the men saw was parachutists raining down from the sky and rolling about on the ground. There was plenty of fighting and they thought it was tremendous fun. The whole battalion volunteered.[11]

A Gurkha paratrooper exits an aircraft. Note the line of open parachutes follow the line of a track on the ground. Military parachutists are dropped at about 800 ft from the ground, not a great deal of time to deploy the reserve if the main parachute fails to open correctly!

Both Gurkha parachute battalions were initially part of 50th Indian Parachute Brigade which deployed to Burma to join General Bill Slim's 14th Army.[12] Of particular note, 153 (Gurkha) Parachute Battalion took part in the Battle of Sangshak from 21 to 26 March 1944, fighting alongside 152 (Indian) Parachute Battalion. Although it had formed, 154 (Gurkha) Parachute Battalion was still in India at the time undergoing training. The large number of young reinforcements joining the battalion meant that its parachute training had had to be delayed and it was therefore not ready for operations until the middle of 1944.[13]

Sangshak was a desperate affair, not least because the majority of the Allied forces arrived at the defensive position only hours before the Japanese assault commenced. After several days of fierce fighting, during which both sides sustained horrendous casualties, the Japanese managed to break through the defences, capturing a church on high ground which dominated the whole position. With their position now untenable, the 50th Indian Parachute Brigade had little option but to withdraw. The war diary of Lieutenant Colonel Paul Hopkinson, then the CO of 152 (Indian) Parachute Battalion, makes particularly sobering reading. He recalls that:[14]

A sketch of the Battle of Sangshak by the war artist Harry Sheldon. Although the 50th Indian Parachute Brigade was eventually over-run, their valiant actions both delayed and blunted the advancing Japanese forces, buying time for the Allies to reinforce their defences at Imphal

> The church position however was still firmly in Japanese hands. Our remaining guns, mortars and the Field Ambulance were now exposed to fire from the church position and had to be moved back towards the centre of the perimeter where there was a little cover. After this there was a lull in the fighting whilst the enemy reorganised and prepared for fresh attacks. At about 1800hrs a message was received in clear over the wireless at Bde HQ from 23 Div which read 'Fight your way out. Go south then west. Air and transport on the look out for you. Good luck our thoughts are with you.'

Although the brigade eventually ceded the position to the Japanese, their determined resistance bought time for 14th Army's defensive positions at Imphal to be strengthened and reinforced with additional troops from the Arakan. In recognition of this, General Slim directed that a Special Order of the Day be produced which praised the Brigade and stated that 'their

The parachute drop at Elephant Point by a composite Gurkha Parachute Battalion on 1 May 1945

staunchness gave the garrison of Imphal the vital time required to adjust their defences.'[15]

A composite Gurkha Parachute Battalion comprised of troops from both 153 and 154 (Gurkha) Parachute Battalions also took part in Operation Dracula, the land, air and sea assault to recapture Rangoon in May 1945. The composite Battalion parachuted onto Elephant Point at the mouth of the Irrawaddy on 1 May 1945 to secure the coastal artillery batteries which dominated the sea approaches to the city.[16]

In November 1945, both 153 and 154 (Gurkha) Parachute Battalions changed their names to become the 2nd and 3rd Battalions (Gurkha) of the Indian Parachute Regiment.[17] On 26 October 1946, the 3rd Battalion

A mortar detachment of Gurkha paratroopers in action at Elephant Point having parachuted in as part of Operation Dracula

Gurkhas from the composite Gurkha Parachute Battalion which secured Elephant Point approach a Japanese bunker. Note the bodies of the dead defenders at the base of the mound

Subedar Major Narbahadur Gurung IDSM and Senior Gurkha Officers from 2nd and 3rd Battalions (Gurkha), Indian Parachute Regiment marching past Major General Chambers, the General Officer Commanding 26th Indian Division. The photograph was taken in Rangoon on 8 May 1945, shortly after the city had been retaken by Allied forces following Operation Dracula

Field Marshal Lord Wavell being introduced to officers and soldiers of 153 (Gurkha) Parachute Battalion

(Gurkha), Indian Parachute Regiment reverted to being the 3rd Battalion of the 7th Gurkhas before being disbanded in November 1946.[18] The 2nd Battalion, Indian Parachute Regiment was eventually disbanded in November 1947.[19]

Both 153 (Gurkha) Parachute Battalion and 154 (Gurkha) Parachute Battalion were real units. However, the 6th (Gurkha) Parachute Brigade was entirely fictional and was created as part of a plan to deceive the Axis powers regarding the true strengths and dispositions of formations that might be deployed to Italy.

On 2 January 1963, the Gurkha Independent Parachute Company was raised by Major L M Phillips of 1/10th Gurkhas from a platoon of 1/10th Gurkhas. Platoons from 1/7th Gurkhas and 2/10th Gurkhas subsequently joined the company which saw active service in Borneo during the confrontation (which is covered in Chapter 8).[20]

Tense faces as members of the Gurkha Independent Parachute Company await the signal to prepare to jump. The reserve parachute is carried on the front and deployed by means of the handle shown in the picture

Paratroopers from the Gurkha Independent Parachute Company preparing to jump. Each soldier's parachute is attached by a 'static line' to a cable which runs the length of the aircraft. As the soldier exits the aircraft, the 'static line' pulls the parachute from a pack on the soldier's back, allowing it to deploy

One of the most interesting tasks undertaken by the company during the Borneo Confrontation was the training of the locally recruited Border Scouts.[21] The scouts, who were intimately familiar with the jungle, were formed to operate along the border with Indonesia, gathering information about enemy activities.[22] The director of operations in Borneo, Major General Walter Walker, understood the value of small groups of determined men operating deep in the jungle – indeed,

A parachute trained company commander in The Royal Gurkha Rifles leads a foot patrol in Helmand Province, Afghanistan. His parachute wings are clearly visible on his right arm above the crossed kukris of the Brigade of Gurkhas.

Soldiers of the Gurkha Independent Parachute Company receiving their maroon berets after having passed their parachute training

The capbadge of the Gurkha Independent Parachute Company. The capbadge is effectively that of the Parachute Regiment set against the red, black and green ribbon of the Brigade of Gurkhas

Gurkhas from the Gurkha Independent Parachute Company during a training jump. The equipment they will need when they land is carried in packs which are strapped to their legs. As he nears the ground, the paratrooper has to release his pack which then hangs on a rope below him, hitting the ground a second or two before he does. Landing with the kit bag still attached to the leg could easily result in broken bones

he famously said that 'I regard 70 troopers of the SAS [Special Air Service] as being as valuable to me as 700 infantry in the role of hearts and minds, border surveillance, early warning, stay behind, and eyes and ears with a sting.'[23,24] With insufficient SAS squadrons available, he directed that both the Guards and Gurkha Independent Parachute Companies were to assume the same role as the SAS, patrolling along the border in the deepest parts of the jungle.[25] Perhaps not surprisingly, the Gurkha paratroopers excelled, quickly establishing a reputation as a formidable fighting force.

The Borneo Confrontation formally ended on 11 August 1966. The Gurkha Independent Parachute Company had played a significant role, demonstrating its utility in a variety of roles, and was not disbanded until 2 December 1971.[26]

When the Gurkha Independent Parachute Company disbanded in 1971, its members returned to their parent battalions and Gurkha military parachuting effectively fell into abeyance for the next two and a half decades.

Soldiers from the Gurkha Independent Parachute Company practise abseiling down from trees in the jungle. This was an essential skill for paratroopers who might be unable to avoid landing high up in the jungle canopy

Gurkha soldiers from the Gurkha Independent Parachute Company on the top deck of a tower about to practise how to exit an aircraft

A Gurkha paratrooper descending towards the jungle in 1968

A Gurkha from the Gurkha Independent Parachute Company practises landing with an instructor. The same technique of rolling on the side of the body with the legs together is still used today to lessen the impact of a parachute landing

However, in 1996, under-manning across the wider British infantry was beginning to have an impact on operational capability and the Brigade of Gurkhas was invited to form a parachute trained Gurkha Reinforcement Company (GRC) for service with the Parachute Regiment. The company, which became C (Gurkha) Company of the Second Battalion of the Parachute Regiment (2 PARA) was an immediate success and deployed on operations with 2 PARA in Bosnia, Macedonia and Afghanistan. The company was eventually disbanded on 25 May 2002.[27]

A Gurkha paratrooper from the Gurkha Independent Parachute Company on patrol in Borneo during the confrontation with Indonesia in the early 1960s

Although there is currently no formalised role for Gurkha paratroopers, there are a large number of soldiers across the Brigade of Gurkhas who are parachute qualified. This is primarily a legacy of the many years that The Royal Gurkha Rifles spent in 5 Airborne Brigade. During this time not only was the GRC formed to support 2 PARA but individual Gurkha officers and soldiers were able to attempt the notoriously difficult All Arms Pre-Parachute Selection course known as 'P Company'. Those who passed the course were then able to complete parachute training, earning the right to wear the coveted wings. The Brigade of Gurkhas also provides a steady stream of volunteers, both soldiers and officers, to serve with Special Forces such as the Special Air Service (SAS). On joining the SAS, all ranks complete military parachute training.

Lieutenant General Sir Hew Pike greets members of the parachute trained Gurkha Reinforcement Company attached to 2 PARA from 1996 to 2002

A group photograph of the Gurkha and British Officers of the 3rd (Gurkha) Battalion, Indian Parachute Regiment (previously 154 (Gurkha) Parachute Battalion). Taken in 1945, the group may be those officers who took part in Operation Dracula, the re-taking of Rangoon in May 1945

Britain is no longer involved in combat operations in Afghanistan and is beginning to reconfigure its forces for contingency deployments across the world. As the French so recently discovered in Mali, such deployments place a premium on the availability of lightly equipped but highly capable airborne forces which are able to respond at short notice. Over the years, Gurkhas have demonstrated that they have a talent for this sort of soldiering. It is therefore perhaps no surprise that, as part of the Army's restructuring, the UK based battalion of The Royal Gurkha Rifles will join 16 Air Assault Brigade in June 2015. On joining the Brigade, the Battalion will adopt the traditional maroon beret of the airborne forces, though whether this move will also mark a formal return to military parachuting remains to be seen. If it does, then there will be ample opportunity to see whether, as Loftus-Tottenham suggested in the 1940s, the Gurkha remains '... probably the best natural parachutist in the world.'[28]

The Late Twentieth Century

1961–2005

Following the successful conclusion of The Emergency, the Gurkha units stationed in Malaya settled into the routine of garrison life. New establishments were introduced for the Gurkha Engineers and the Gurkha Signals. The Gurkha Army Service Corps, which had been formed in July 1958, was also expanded to four squadrons.[1] Although Malaya was the Gurkhas' base location, units were also stationed in Hong Kong and Singapore. But the peace did not last long. At 0200hrs on Saturday 8 December 1962, insurgents belonging to an organisation called

Soldiers from 1/2nd Gurkhas deploying to Brunei from their base location in Singapore. The Brunei Revolt started at 0200 hours on Saturday 8 December 1962. 1/2nd Gurkhas arrived in Brunei that evening to begin operations against the insurgents

the North Kalimantan National Army (Malay abbreviation TNKU) launched a rebellion in the British Protectorate of Brunei.[2] Their intention was to take the Sultan of Brunei hostage, capture the main police stations and seize the important oilfield in Seria.[3]

Although the revolt caught many by surprise, it had been brewing for some time.[4] Sarawak, North Borneo (which changed its name to Sabah in 1963) and Brunei lie on the northern coast of the tropical island of Borneo. To the south lies the Indonesian state of Kalimantan. The British intention was to grant independence to North Borneo, Sarawak and Singapore in order that they could combine with Malaya and Brunei to form a Federation of Malaysia.[5] But President Sukarno of Indonesia had other ideas. He wanted the whole of Borneo, including Sarawak, Brunei and North Borneo, to come under Indonesian leadership.[6] In September 1962, Brunei held elections for 17 of the 35 seats in Brunei's Legislative Council. All were won by a political party called the *Partai Raayatt* (People's Party). The remaining 18 seats were directly nominated by the sultan. The *Partai Raayatt* broadly supported Sukarno's aims but was outvoted when the issue was considered by the Legislative Council. The *Partai Raayatt*'s response to this loss was to mobilise its action arm, the TNKU, and to start an armed revolt against the sultan.

The British responded to the Brunei Revolt by deploying a mixed force of Gurkhas, British Infantry and Royal Marines. 1/2nd Gurkhas, who

Belvedere helicopters were used during the Brunei Revolt to deploy troops throughout the area of operations

A soldier from 1/2nd Gurkhas stands guard in the town of Tutong in Brunei

Soldiers from 1/2nd Gurkhas guarding rebel prisoners being held in a cinema in Brunei

were based in Singapore, were the first to arrive with D Company landing in darkness on the night of the 8 December 1962.[7] After a number of 'sharp skirmishes', they managed to secure Brunei Town.[8] The remainder of the force arrived over the next few days, deploying to the south and north of the country to secure key installations and rescue hostages taken by the insurgents. The situation remained confused for a number of days but, within a week, some 3000 of the 4000 rebels who had participated in the revolt had either been captured or had surrendered.[9]

By 17 December 1962, order had largely been restored and the British force had turned its attention to tracking down the remaining insurgents who had fled into the jungle. These included Yassin Affendi, the TNKU's military commander.[10] In January 1963, 1/2nd Gurkhas were replaced by 2/7th and it was this battalion's B Company which, on 18 May 1963, brought the Brunei Revolt to a conclusion by capturing Affendi.[11]

During the revolt, L Company Royal Marines had conducted a daring operation to rescue hostages being held in the town of Limbang. The commander of L Company was Jeremy Moore. Twenty one years later, he would again fight alongside Gurkhas as the commander of land forces in the Falklands. There is little doubt that the prompt deployment of 1/2nd Gurkhas made a significant contribution to Britain's successful response to the Brunei Revolt. But it came at a cost: the battalion lost two killed, including a British officer, and 17 wounded; the insurgents lost 65 killed.[12] A further 783 rebels were either captured or surrendered.[13]

Major General Walter Walker,[14] the commander of 17th Gurkha Division, arrived in Brunei on 19 December 1962 to take overall command as commander land forces, Borneo.[15] He believed that the political leader of the TNKU, 'Sheikh' Azahari, was working for President

Although helicopters were used to great effect throughout the Borneo Confrontation, the nature of the terrain meant that soldiers had to patrol on foot as they played a deadly game of hide and seek against the Indonesian forces

Taken in 1964, this photograph shows mortars of the 2nd Gurkhas firing from a trench during the Borneo Confrontation

Sukarno and that the Brunei Revolt was 'merely the prelude to something much bigger.'[16] He was right. In April 1963, a group of Indonesian Border Terrorists (IBT) crossed the border into the British colony of Sarawak and attacked a police outpost.[17] The IBT were well trained and were led by regular Indonesian soldiers.[18] General Walker, whose appointment expanded when he became the director of operations in Borneo in April 1963, noted that they were 'very bright, a first-class enemy.'[19] The involvement of the IBT made the Borneo Confrontation (as the campaign was to be known) significantly different from the British campaign in Malaya. In the latter, the British had been fighting an internal enemy but in Borneo the British were primarily fighting an external threat.[20] As the campaign developed, it became increasingly apparent that greater numbers of Indonesia's regular army the Tentara Nasional Indonesia (TNI), were being committed to operations against the British.[21] An internal threat did exist in both Sarawak and North Borneo in the form of the Clandestine Communist Organisation (CCO) but its role was largely restricted to providing the IBTs and Indonesian troops with information and material support.

Initially, the British focused their efforts on the 1000 mile border with Indonesia. Working with the Special Air Service (SAS), they tried to interdict Indonesian patrols as they crossed the border into Sarawak and Sabah. But the dense jungle nature of the border meant that this was incredibly difficult, even with the support of the local population. There were numerous instances of large parties of IBT and TNI

The Gurkha engineers played a key role in Borneo, building bridges and constructing roads through the densest of jungles

Soldiers from 69 Gurkha Field Squadron (which was formed in April 1962) unload a light tractor which has been air dropped into a work site in the jungle

crossing the border and attacking outposts. One such incident occurred in December 1963 in a place called Kalabakan in Sabah.[22] An IBT and TNI raiding party of about 200 attacked a small garrison of the Royal Malay Regiment, killing eight and wounding 19 without a single casualty to themselves.[23] In response, 1/10th Gurkhas spent the next two months tracking the attackers down, steadily eliminating them in a series of jungle ambushes as they tried to get back across the border with Indonesia.[24] Such operations were typical of the next few years with Gurkhas and Indonesian combatants hunting each other in the dense jungle that characterises Borneo. It was a deadly game of hide-and-seek. Although the Gurkhas excelled at jungle warfare, the Indonesians were also extremely capable in this environment, particularly their elite paratroopers and marines.

General Walker realised that unless he could cross the border and take the fight to the Indonesians, he would be unable to seize the initiative.

The British made widespread use of helicopters in Borneo to support their ground operations with Gurkha units forming a particularly close relationship with the Royal Navy's Commando helicopter squadrons. The pilots of their Wessex helicopters became known as 'Junglies', a nickname that endures to this day.

He sought permission to do this and, eventually, the British Government agreed. In May 1964, the first 'Claret' operations as they were known were born.[25] The ability to operate across the border changed the dynamic of the campaign but it was not without its challenges. Helicopters, for example, were not allowed to cross the border and this meant that patrols had to be inserted by foot.[26] Water, food, ammunition, radios, first aid equipment and spare clothing all had to be carried for each ten day operation, a significant task given the nature of the terrain and the climate. 'Claret' operations fell into two basic categories: either an ambush on a route known to be used by the enemy or an attack on a known base.[27] Numerous 'Claret' operations of both types were carried out. Although hazardous, these operations, which remained a well-kept secret for a decade after the end of the confrontation,[28] had immediate impact, taking the fight to the enemy and restricting his freedom of manoeuvre.

Helicopters were widely used to insert Gurkhas into defensive positions on the high ground that dominated the border with Indonesia. Inevitably, landing in such conditions could be precarious. In one accident, a Regimental Medical Officer had to use a penknife to amputate the arm of Major 'Birdie' Smith in order that he could be freed from an overturned helicopter[29]

On one 'Claret' operation, a company from 2/10th Gurkhas, commanded by Captain Kit Maunsell, was tasked with infiltrating two miles across the border to seize an enemy prisoner. On 21 November 1965, they were close to the enemy camp having approached along a knife ridge when a machine gun opened fire on them. One of the Gurkhas in Lance Corporal Rambahadur Limbu's section was hit. Rambahadur ran forwards, reaching the enemy trench in seconds, and killed the sentry, gaining a foothold in the enemy position.[30] Running into the open to try and communicate with his platoon commander, Rambahadur noticed that two of his men were seriously wounded. He tried to crawl towards them but the weight of enemy fire was too great. He realised that the only way he could get to them would be to dash through the hail of bullets, hoping that his speed would '... give him the cover which the ground could not.'[31] The citation for the Victoria Cross that he was awarded for his actions explains what happened next:

> Rushing forward he hurled himself on the ground beside one of the wounded and calling for support from two light machine guns which had now come up on his right in support he picked up the man and carried him to safety out of the line of fire. Without hesitation he immediately returned to the top of the hill determined to repeat his self imposed task of saving those for whom he felt personally responsible ... Picking him up and unable now to seek cover he carried him back as fast as he could through the hail of enemy bullets.[32]

As the citation goes on to note 'that he was able to achieve what he did against such overwhelming odds without being hit is miraculous.'[33]

Not all the fighting took place on Borneo. In September 1964, Indonesian paratroopers parachuted into the jungle around the town of Labis in mainland Malaya.[35] 1/10th Gurkhas, who were on rest and recuperation (R&R) leave from Borneo, were despatched to deal with the attacking force. After a number of spirited engagements, 1/10th eventually succeeded in capturing or killing all 96 of the enemy paratroopers but,

Lance Corporal Rambahadur Limbu of 2/10th Gurkhas who was awarded a Victoria Cross for his actions on 21 November 1965 during a secret cross border operation to try and capture an Indonesian prisoner

Gurkhas embarking in RN helicopters from a carrier deck. The Director of Operations in Borneo, Major General Walter Walker, insisted on high levels of 'Jointmanship',[34] with the Army, Navy and Air Force all working together to achieve a common purpose

In September 1964, Indonesian paratroopers landed near the town of Labis in Malaya. 1/10th Gurkhas, who were on rest and recuperation from operations in Borneo, were despatched to deal with the attackers. This photograph shows Gurkhas from 1/10th Gurkhas with some of the captured paratroopers

again, it came at a cost with 1/10th losing a British officer and a Gurkha lance corporal.[36]

The Indonesians were not the only ones relying on paratroopers to achieve decisive effect. In 1961, a Gurkha Independent Parachute Company was formed with men drawn from the 10th and 7th Gurkhas.[37] The company deployed to Borneo and, following the example of the SAS, conducted five-man patrols along the border with Indonesia.[38]

On 11 August 1966 the Borneo Confrontation came to an end with the signing of a peace agreement in Jakarta.[39] In all, some 17,000 British, Gurkha and Commonwealth troops had been deployed in Sarawak, Sabah and Brunei during the confrontation.[40] Britain lost 114 killed (of which 43 were Gurkhas) and 181 wounded; the Indonesians fared far worse with an estimated 1583 casualties (of which 771 were captured).[41] For three years and nine months, Britain's Gurkhas had been right in the thick of the action, carrying the lion's share of the operational burden in Borneo. In doing this, not only did they directly contribute to the end of the conflict but they also reinforced their reputation as the British Army's most capable jungle fighters, a reputation that endures today.

A Gurkha from the Gurkha Independent Parachute Company which was formed in 1961 practising how to abseil down from the jungle canopy

Support Company of 2/2nd Gurkhas about to leave Lundu in Sarawak having handed over their responsibilities to 42 Commando in late December 1945

The rivers in Borneo are used as arterial transport routes. During the confrontation, both British and Indonesian forces used them to move troops and supplies. In August 1965, Support Company of 2/2nd Gurkhas carried out a highly successful boat ambush on the River Sentimo, killing 12 Indonesians who were travelling in an assault boat

Soldiers from the 7th Gurkhas manning a border post on the Hong Kong-Sino Border. From May to June 1967, the Gurkhas were heavily involved in internal security operations in support of the Royal Hong Kong Police as they sought to control widespread rioting

Barely six months after the ending of the confrontation, cuts imposed by the Ministry of Defence meant that the number of Gurkhas was drastically reduced from 15,000 to 7000. This entailed many fine soldiers being discharged with no pensions and very small gratuities. Realising that this would add to the number of ex-British Gurkhas living in conditions of privation in Nepal, The Gurkha Welfare Trust was formed with generous donations from business and commerce throughout the UK and Far East.

In May and June 1967, rioting broke out in Hong Kong, fuelled by the communist authorities in mainland China. The Gurkhas responded, supporting the police as they sought to restore order. On 8 July 1967, the situation deteriorated further when Chinese rioters attacked the police station in the border town of Sha Tau Kok.[42] As the Hong Kong Police responded, a Chinese machine gun opened fire from across the border, killing two and wounding five.[43] 1/10th Gurkhas deployed, eventually recovering the casualties but this was only the start. In the weeks that followed, the Gurkhas were to find themselves fully committed to riot control in Hong Kong's congested urban areas as well as on the border.

By 1971, the Gurkhas had withdrawn from Malaysia and were established in their new base location in Hong Kong. In addition, in 1972 the Ministry of Defence began stationing a Gurkha battalion in the

Soldiers of the 10th Gurkhas arrive at the border town of Sha Tau Kok on 8 July 1967 to support the Royal Hong Kong Police. Rioters had stormed the town's police station earlier in the day. As the police responded, a machine gun from across the Chinese border opened fire, killing two and wounding five. 1/10th Gurkhas eventually managed to recover the casualties

Soldiers of the 10th Gurkhas deplaning in Cyprus in 1974. The battalion was deployed at short notice to help protect the Sovereign Base Areas when Turkish forces invaded the island

UK. Gurkha battalions then rotated between Hong Kong, the UK and Brunei. The latter posting was at the request of the Sultan.

In 1974, 10th Gurkhas were the UK Gurkha Battalion when they were hastily deployed to Cyprus to help protect the UK Sovereign Base Areas following the Turkish invasion of the island.[44] The battalion acquitted itself particularly well during this short but tense operation, earning accolades from both the commander British Forces in Cyprus and the British high commissioner.

In 1975, the military garrison in Hong Kong was deployed to help stem the tide of illegal immigrants trying to escape from communist China into Hong Kong.[45] By 1978, Gurkha units were spending the majority of their time patrolling the border and catching illegal immigrants (IIs),[46] a task which was made easier after the Hong Kong Government constructed a fence, topped with razor wire, which ran the length of the border. The author, who joined his battalion in Hong Kong in 1987, spent much of the first few years of his Army service patrolling the border with his platoon of Gurkhas from D Company 1/2nd Gurkhas. Whilst a benign experience compared to what today's young officers

A soldier from 10th Gurkhas carrying weapons recovered from Cypriot refugees following the Turkish invasion of the island in 1974

Gurkhas from the 10th Gurkha's Reconnaissance Platoon patrolling in Ferret armoured cars in Cyprus in 1974

Soldiers from the 7th Gurkhas escorting illegal immigrants (II) captured trying to enter Hong Kong from mainland China. The IIs were handed back to China on a daily basis

As well as trying to enter Hong Kong by land, Illegal Immigrants would often try and swim across the many bays and inlets that characterised the border with mainland China. Gurkhas routinely patrolled these, all too often recovering the bodies of those who had sadly drowned trying to reach the bright lights of Hong Kong

have to contend with in Afghanistan, it was not without its challenges. One of the most frustrating of these was the mosquitoes which, particularly in the coastal marshes, appeared to have developed immunity to insect repellent!

The task of patrolling the border and capturing IIs became routine but there were a number of occasions when Gurkhas demonstrated the courage and determination for which they have become rightly famous. One of these occurred on the night of 28 May 1979. Lance Corporal Aimansing Limbu of 7th Gurkhas was in charge of a four-man anti-II patrol near the Chinese border. His team were lying in ambush on a beach, waiting to intercept IIs. The sea was rough and visibility was poor. At 0200hrs, the Gurkhas spotted a raft about 100 metres from the shore. A lone individual was then spotted leaving the water about 50 metres from the ambush position. Leaving the rest of his team to observe the raft, Lance Corporal Aimansing Limbu went to apprehend the lone II. As he tried to arrest the individual, he was attacked by two others who knocked him to the ground. All three of the IIs then jumped

Soldiers from the 6th Gurkhas capture an illegal immigrant (II) attempting to climb over the border fence that separated Hong Kong from China. Over ten foot tall and topped by double strands of razor wire, it was remarkable how quickly an II could scale the fence and disappear into the undergrowth

into the sea and started to swim back towards the raft. Lance Corporal Aimansing Limbu quickly grabbed a rubber ring discarded by one of the IIs and jumped in after them. Two of the IIs attempted to drown Aimansing but were 'discouraged by blows from his baton and were rendered capable of only clinging onto the raft.'[47] Aimansing then managed to climb onto the raft and, despite 'kicks, blows and bites',[48] managed to subdue the six men on the raft, forcing them to paddle to the shore where all eight were apprehended by the rest of Aimansing's patrol. Lance Corporal Aimansing was awarded the Queen's Gallantry Medal for his actions, the first ever awarded to a Gurkha.[49]

Gurkhas continued to patrol the border until just before Hong Kong was handed over to the Chinese Government in 1997. It was in many ways a thankless task and the Gurkhas deserve considerable credit for patrolling the border so diligently in the decades leading up to the handover.

On 2 April 1982, Argentina invaded the Falkland Islands. At the time, 1/7th Gurkhas were the resident UK Gurkha battalion and were part of 5 Infantry Brigade, a high readiness formation specifically set up for 'out of area' operations.[50] On 12 May 1982,[51] the battalion departed Southampton on the requisitioned cruise liner Queen Elizabeth II and began the long journey to the South Atlantic as part of the British Task Force assembled to liberate the Falklands. The two parachute battalions which ordinarily formed part of 5 Infantry Brigade had been detached to augment 3 Commando Brigade.[52] Their place in 5 Infantry Brigade was eventually taken by the Scots Guards and the Welsh Guards.[53]

By the time 1/7th Gurkhas arrived in the Falklands on 1 June 1982, 3 Commando Brigade had landed and cleared Darwin and Goose Green. 1/7th Gurkhas were directed to occupy these key locations, as well as to look after 500 Argentine prisoners that had been taken by the advancing 3 Commando Brigade. The Gurkhas were also tasked with clearing the area to the south (known as Lafonia) of small pockets of Argentine resistance. They executed this latter task with their customary enthusiasm, at one point using a kukri to frighten three Argentine soldiers into submission![54]

There is no doubt that the units of 3 Commando Brigade did the lion's share of the fighting but, in the early hours of 14 June 1982, 1/7th

A Gurkha from 1/7th Gurkhas mans a radio set next to a command post constructed out of ammunition boxes

Soldiers from 1/7th Gurkhas manning a command post shortly before their assault on Mount William. Despite the freezing conditions and the imminent prospect of action, the soldiers continued to smile

Soldiers from 1/7th Gurkhas with a captured Argentinian anti-aircraft gun. The battalion deployed as part of 5 Infantry Brigade and took part in the final battle for Port Stanley

Gurkhas and the Scots Guards took part in the final attack to seize the high ground that dominated Port Stanley. The Scots Guards' objective was a feature called Mount Tumbledown. The Gurkhas' objective was a neighbouring feature called Mount William.[55] In the event, the Argentine defenders surrendered to the Scots Guards before the Gurkha attack could gain momentum, reportedly because they were afraid of the Gurkhas![56]

The 7th Gurkhas were the first to admit that, compared to other units in the Task Force, they had little actual involvement with the enemy.[57] But they did what was asked of them and they did it well. Moreover, had they been asked to do more, there is little doubt that they would have risen to the challenge with their usual alacrity. The battalion returned to UK 95 days after their deployment, receiving a hero's welcome.

The 1990s saw elements of the Brigade of Gurkhas deployed on a number of operations. During the First Gulf War, for example, the Band of The Brigade of Gurkhas deployed to Kuwait with 28 (Ambulance) Squadron of the Gurkha Transport Regiment (GTR). Together, they formed the Gurkha Ambulance Group. As Brigadier Christopher Bullock notes, this was the first time that Gurkhas had served on operations in the desert since the Second World War.[58] At the time, the band was capbadged to the 2nd Gurkhas and was accommodated within the First Battalion's lines. The author, who was the adjutant of 1/2nd Gurkhas

1/7th Gurkhas return to the UK aboard SS Uganda, 5 August 1982. The Battalion deployed as part of the Task Force sent by Prime Minister Margaret Thatcher to liberate the Falkland Islands following their invasion by Argentine forces on 2 April 1982

Soldiers from 1/7 th Gurkhas marching through the town of Fleet in Hampshire after their return from the Falkland Islands. Fleet was the nearest town to the resident UK Gurkha battalion's home base of Church Crookham

Gurkhas from 69 Gurkha Field Squadron of The Queen's Gurkha Engineers building a bridge in Gracanicia, Bosnia in 1996

over this period, well remembers the band returning from the Gulf with medals that were the envy of the rest of the battalion! Notably, the band rebadged to The Royal Gurkha Rifles on 1 July 1994 when, faced with reductions in the size of the Gurkha Infantry, the 2nd, 6th, 7th and 10th Gurkhas amalgamated to form a new regiment. Also in the 1990s, the Gurkha Transport Regiment deployed a squadron to Cyprus to take part in a UN peacekeeping mission.

The last Gurkha unit left Hong Kong in 1997 just prior to the colony being handed over to the Chinese Government. Now based predominantly in the UK, the brigade was as busy as ever as Gurkhas continued to deploy on operations. The Gurkha-manned 28 Transport Squadron

Soldiers from 3RGR on operations in Bosnia in 1996 as part of the NATO Intervention Force (IFOR)

was soon deployed to the former Republic of Yugoslavia as part of the UN Protection Force (UNPROFOR), quickly followed by other Gurkha units as NATO's role changed to one of active intervention (IFOR) in an attempt to separate the warring factions.[59]

In June 1999, the First Battalion of The Royal Gurkha Rifles (1RGR) deployed to Macedonia as part of 5 Airborne Brigade. After a tense few days, 1RGR crossed the border into Kosovo as part of the Kosovo Intervention Force (KFOR), the route having been cleared of mines by The Queen's Gurkha Engineers.[60] Their role was to keep the peace between the Serbs and the Kosovans, a delicate task made all the more demanding because of the ubiquitous presence of the media. Troops from The Queen's Own Gurkha Logistic Regiment and Queen's Gurkha Signals also served in Kosovo.

In September 1999, the Second Battalion of The Royal Gurkha Rifles (2RGR) deployed from Brunei to East Timor as part of a joint operation with the Australians. The intervention force (known as INTERFET) was tasked with trying to restore order as rebels supported by Indonesia sought to destabilise the country to prevent it gaining independence. Further deployments followed with 1RGR flying to Sierra Leone in 2001 to separate warring factions and help develop the country's armed forces as part of Operation Silkman.[61]

For the next few years, all the Gurkha units were kept busy deploying troops to sustain the UK's operations in Kosovo and Bosnia. In addition, the Gurkhas returned to operations in Afghanistan after an absence of

Gurkhas from 1RGR practise riot control techniques in Bosnia, December 2005

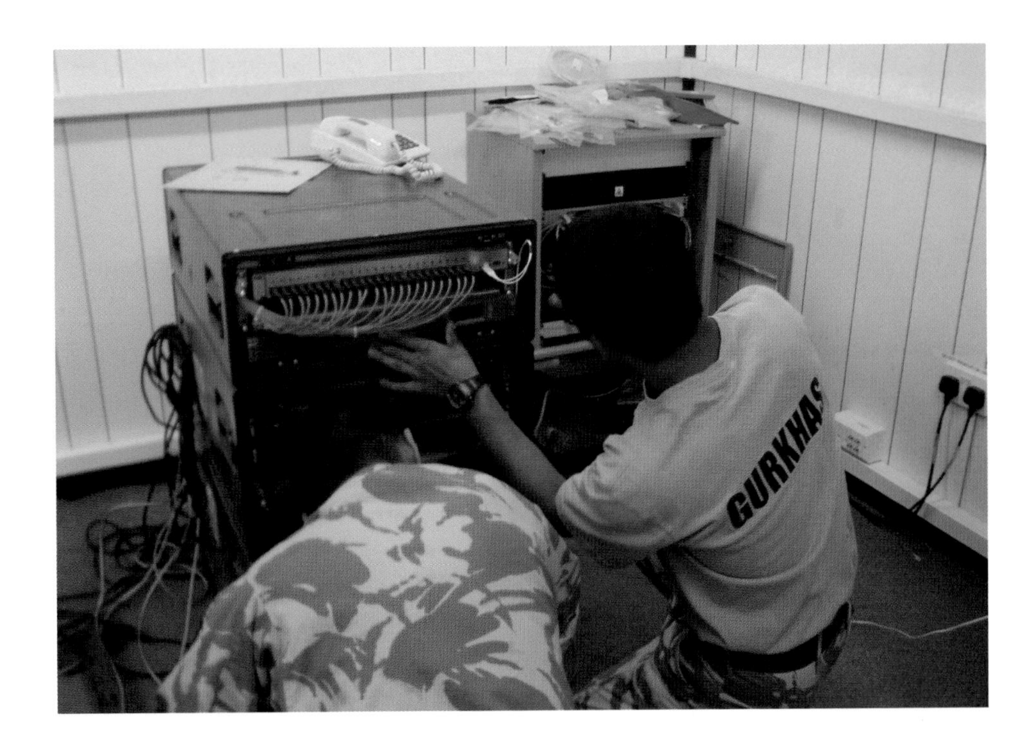

Gurkha signallers resolving communications problems in Iraq on Operation Telic. Gurkha signallers complete the same technically demanding courses as their British counterparts and are equally as capable at handling today's advanced communications systems

over half a century. This is covered in detail in the next chapter. In 2003, a company of Gurkhas deployed with the 1st Battalion of the Royal Irish Regiment on Operation Telic, the invasion of Iraq.[62]

The Queen's Gurkha Engineers also deployed to Iraq with 69 Gurkha Field Squadron playing a key role in the construction of essential infrastructure. Troops from both the Queen's Gurkha Signals and The Queen's Own Gurkha Logistic Regiment also took part in Operation Telic as routine deployments to Iraq became a feature of life in the supporting arms.

As the images and narrative in this chapter have shown, the latter half of the 20th century was a remarkably busy period for all Britain's Gurkha units. Whether fighting insurgents in the jungles of Brunei and Borneo or keeping the peace between warring factions in the Balkans, the Gurkhas demonstrated complete professionalism and an ability to adapt to the circumstances. That they were able to master their environment so quickly, setting the pace in delivering operational effect, may have surprised those who thought that Gurkhas would struggle to adapt to contemporary operations. But whilst the nature of warfare may have changed since Gurkhas were first recruited in 1815, the characteristics that made them effective then continue to make them successful now. Self-discipline, humility, physical resilience, courage and determination are as important in today's conflicts as they were 200 years ago. Fortunately, the Gurkha, then as now, has them in spades.

The Victoria Crosses of the Brigade of Gurkhas

Although the Victoria Cross was brought into being by Royal Warrant on 29 January 1856, it was not until 1911, on the occasion of the visit of King George V to India, that Gurkhas became eligible for the award.[1] To date, officers and soldiers of the Brigade of Gurkhas have been awarded a total of 26 Victoria Crosses. Of these, 13 have been awarded to British Officers and 13 to Gurkha officers and soldiers.[2]

The reverse of each Victoria Cross is engraved with the recipient's Army number, name, unit and the date of the action for which the medal was awarded. Rifleman Kulbir Thapa, whose medal is shown here, was the first Gurkha soldier to be awarded a Victoria Cross. He received it for his actions at the Battle of Loos on 25 September 1915

Recipients of the Victoria Cross

Name	Regiment	Year
Lieutenant J A Tytler	1GR	1858
Major D Macintyre	2GR	1872
Captain G N Channer	1GR	1875
Captain J Cook	5GR	1878
Captain R K Ridgeway	8GR	1879
Lieutenant C J W Grant	8GR	1891
Lieutenant G H Boisragon	5GR	1891
Lieutenant J Manners-Smith	5GR	1891
Captain W G Walker	4GR	1903
Lieutenant J D Grant	8GR	1904
Rifleman Kulbir Thapa	3GR	1915
Major G C Wheeler	9GR	1917
Rifleman Karanbahadur Rana	3GR	1918
Subedar Lalbahadur Thapa	2GR	1943
Havildar Gaje Ghale	5RGR	1943
Lieutenant (Acting Captain) M Allmand	6GR	1944
Rifleman Ganju Lama MM	7GR	1944
Rifleman Tulbahadur Pun	6GR	1944
Jemedar (Acting Subedar) Netrabahadur Thapa	5RGR	1944
Rifleman (Acting Naik) Agansing Rai	5RGR	1944
Captain (Temporary Major) F G Blaker MC	9GR	1944
Rifleman Sherbahadur Thapa	9GR	1944
Rifleman Thaman Gurung	5RGR	1944
Rifleman Bhanbhagta Gurung	2GR	1945
Rifleman Lachhiman Gurung	8GR	1945
Lance Corporal Rambahadur Limbu	10GR	1965

Of the 26 Victoria Crosses listed above, the circumstances leading to the award of 20 of them are described in detail in the chapters of this book. But those relating to six of them are not mentioned. As these medals were awarded for extreme acts of valour, it is worth summarising the circumstances that led to their recipients being recommended for the medal. The majority of the detail in this section is taken from Maurice Biggs's excellent book *The Story of Gurkha VCs*, which is available from the Gurkha Museum.

Captain George Nicholas Channer VC

George Nicholas Channer who was awarded a Victoria Cross for his actions as a captain whilst serving with the 1st Goorkha Regiment (later the 1st Gurkha Rifles) during the Perak Expedition of 1875–1876

George Channer was born in India in January 1843 and was the son of a Colonel George Girdwood Channer and his wife Susan.[3] He was commissioned in 1859 and served with a number of different units before being attached to the 1st Gurkhas (then known as the 1st Goorkha Regiment) on 6 March 1873. As explained in Chapter 3, the Indian Army spent much of its time in the late 1800s fighting rebellious tribesmen on the frontiers of British India. When Britain's wider interests were threatened, units from the Indian Army occasionally ventured further afield.

On 2 November 1875, Mr James Wheeler Woodford Birch, the British resident in the Malayan state of Perak, was murdered whilst bathing in the Perak River. His attackers were followers of Sultan Abdullah of Perak. The sultan had signed a treaty with the British in 1874 but this had been done to help him consolidate his position as ruler rather than because he had any deep affection for the colonial power. In the months prior to Birch's death, the sultan had let it be known

Officers' quarters in Campong Boyah during the Perak Expedition of 1875–1876. This image appeared in the *London Illustrated News* on 26 February 1876

Soldiers from the 1st Goorkha Regiment storming the fort at the Bukit Putas Pass during the Perak Expedition of 1875–1876. The assault was led by Captain G N Channer who was awarded a Victoria Cross for his actions

that he disliked the British and that he did not want them in Perak. Local Malay rajahs and chiefs therefore began ignoring British advice and it was this civil disobedience which escalated and eventually culminated in Birch's murder.

At the time, the senior British authority in Malaya was the governor of the so-called Straits Settlements of Penang, Malacca and Singapore. Fearing that the disobedience might spread throughout Malaya, he requested that reinforcements be sent from India to restore order.

The force from India included the 1st Goorkha Regiment. They joined additional troops from Malacca and Hong Kong to form three fighting groups. The first of these, which included one company of the 1st Goorkhas, was the Perak Field Force. On 8 December 1875, this set off up the River Perak to try and capture withdrawing rebel leaders. The second group, which was known as the Larut Field Force and which included about 400 men from the 1st Goorkhas, was despatched up the River Larut to disarm a particularly anti-British village. The third force, which was known as the Malacca Column, was despatched south of Perak and deep into what is now the state of Negeri Sembilan. This force included some 250 men from the 1st Goorkhas. Its mission was to 'remove obstacles' to the rule of a particular chief who was a supporter of the British.[4] This involved destroying the defended stockades of a rival chief and dispersing his supporters but, to gain access to these stockades, the Malacca Column first needed to capture the key Bukit Putas Pass. Captain Channer, who was commanding a detachment of the 1st Gurkhas, was ordered forward to reconnoitre a fort which dominated the approaches to the Pass. Subsequently promoted to brevet major, the citation for Captain Channer's Victoria Cross picks up the story:

> For having, with the greatest gallantry, been the first to jump into the Enemy's Stockade, to which he had been despatched with a small party of the 1st Ghoorka Light Infantry, on the afternoon of the 20th December, 1875, by the Officer commanding the Malacca Column, to procure intelligence as to its strength, position, etcetera.
>
> Major Channer got completely in the rear of the Enemy's position, and finding himself so close that he could hear the voices of the men inside, who were cooking at the time, and keeping no look out, he beckoned to his men, and the whole party stole quietly forward to within a few paces of the Stockade. On jumping in, he shot the first

A Group photo taken in 1870 of Gurkha officers of the 1st Goorkha Regiment (later to become the 1st Gurkha Rifles)

> man dead with his revolver, and his party then came up, and entered the Stockade, which was of a most formidable nature, surrounded by a bamboo palisade; about seven yards within was a loghouse, loop-holed, with two narrow entrances, and trees laid latitudinally, to the thickness of two feet.
>
> The Officer commanding reports that if Major Channer, by his foresight, coolness, and intrepidity, had not taken this Stockade, a great loss of life must have occurred, as from the fact of his being unable to bring guns to bear on it, from the steepness of the hill, and the density of the jungle, it must have been taken at the point of the bayonet.[5]

By March 1876, the Perak War was over. A number of the insurgent leaders were hanged and the rebellious sultans and chiefs reluctantly accepted British rule. Channer himself went on to have a highly successful military career, rising through the ranks in short order. Interestingly, he commanded the 1st Brigade during the Black Mountain Expedition of 1888 which is described in detail in Chapter 3.[6] Promoted to lieutenant general on 9 November 1896, he died on 13 December 1905 at the age of 62.

Lieutenant Richard Kirby Ridgeway VC

Richard Ridgeway was born in Ireland on 18 August 1848. He served in a number of Indian Army units before being posted to the 44th (Sylhet) Light Infantry (which later became the 1st Battalion, 8th Gurkha Rifles) on 3 July 1874. In 1879, tribesmen in the Naga Hills in the north-east of India were causing trouble and an expedition was launched to impose order. The force included both the 44th and the 43rd (which later became the 2nd Battalion, 8th Gurkha Rifles). The Naga stronghold was the village of Konoma. The defences were based on a series of terraces, each of which had strong stone walls interspersed with towers which dominated the approaches. On the 22 November 1879, the 44th mounted an attack against the stronghold. Although they succeeded in capturing part of the outer defences, they were initially unable to breech the central stockade. Lieutenant Ridgeway led two gallant assaults against this key position which was defended by 'several thousand well-armed Nagas.'[7] The first assault was unsuccessful. The citation

Richard Kirby Ridgeway who, as a lieutenant with the 44th (Sylhet) Light Infantry (which later became the 1st Battalion, 8th Gurkha Rifles), was awarded a Victoria Cross for his actions during the Naga Hills Campaign of 1879–1880

A Gurkha officer of the 44th (Sylhet) Light Infantry (which later became the 1st Battalion, 8th Gurkha Rifles) circa 1880

Lieutenant Ridgeway was promoted to Captain on 8 January 1880. In 1891, he took command of his regiment (which had been renamed the 44th Gurkha (Rifle) Regiment of Bengal Infantry). He commanded the regiment during the Manipur Campaign later that year and was mentioned in despatches. This picture, taken from the *Illustrated London News*, shows the regiment deploying at the start of the Manipur Campaign of 1891

The 44th (Sylhet) Light Infantry (later 1st Battalion, 8th Gurkha Rifles) taken in 1884

for the Victoria Cross he was subsequently awarded describes what happened in the second assault:

> For conspicuous gallantry throughout the attack on Konoma, on the 22nd November, 1879, more especially in the final assault, when, under heavy fire from the enemy, he rushed up to a barricade and attempted to tear down the planking surrounding it, to enable him to effect an entrance, in which act he received a very severe rifle shot wound in the left shoulder.

Having torn down the planking, Lieutenant Ridgeway remained in his position in full view of the enemy until his men had climbed through the gap and into the stockade. The Nagas were eventually defeated. Ridgeway subsequently took part in the Manipur Expedition of 1891 and the Tirah Expedition of 1897, both of which are described in detail in Chapter 3. Ridgeway eventually retired as a Colonel, passing away on 11 October 1924 at the age of 76.

Lieutenant Guy Hudleston Boisragon VC

Guy Boisragon was born in India on 5 November 1864. His father was Major General H F M Boisragon, a distinguished soldier who in 1858 had raised the 25th Punjab Infantry or Huzara Goorkha Battalion (which later became the 5th Royal Gurkha Rifles (Frontier Force)). On 7 April 1887, Boisragon followed in his father's footsteps by joining the 1st Battalion, 5th

Guy Hudleston Boisragon who, as a lieutenant serving with the 5th Goorkha Regiment (later to become the 5th Gurkha Rifles (Frontier Force)) was awarded a Victoria Cross for his actions with the Hunza-Nagar Field Force on 2 December 1891

Goorkha Regiment, The Hazara Goorkha Battalion. His early years were spent campaigning on the North West Frontier with his battalion. In September 1891, he was part of an expedition which was despatched to deal with disobedient Hunza and Nagar tribesmen occupying territory on the Hunza River in Kashmir. The small expedition consisted of some 200 men from 1/5th Gurkhas and a section of the Hazara Mountain Battery. Additional troops were eventually attached to the expedition and it became the Hunza-Nagar Field Force with a strength of some 1130 rifles.

The field force's immediate objective was a fort at a place called Nilt. The fort was strongly built with stone walls which were 14 ft high and 8 ft thick. The fort sat on a precipice over the River Hunza which made access particularly difficult. Lieutenant Boisragon, supported by a group of Bengal sappers under the command of Captain Aylmer, was ordered to try and break through the main gate of the fort. Eventually, the sappers succeeded in blowing the charge and a small number of Gurkhas stormed into the fort. But there were insufficient of them to take it and so Boisragon ran back under constant and heavy enemy fire to collect reinforcements. The citation for the Victoria Cross that Boisragon was awarded for 'conspicuous bravery in the assault and capture of the Nilt Fort on 2nd December, 1891' describes what happened next:

> This Officer led the assault with dash and determination, and forced his way through difficult obstacles to the inner gate, when he returned for reinforcements, moving intrepidly to and fro under a heavy cross-fire until he had collected sufficient men to relieve the hardly pressed storming party and drive the enemy from the fort.

Officers serving with the 5th Gurkhas in November 1904. Note that Major Boisragon VC is shown top right

The officers of the 5th Gurkha Rifles (Frontier Force) taken in 1904. Guy Boisragon is on the back row, seventh from the left. Unfortunately, his Victoria Cross is obscured by the head of the officer in front!

The fort was eventually secured. Its capture marked the beginning of the end for the disobedient Hunza and Nagar tribesmen. In the subsequent battle for the Thol Cliffs on 20 December 1891, another officer of the 5th Gurkhas, Lieutenant J Manners-Smith, was also awarded the Victoria Cross. Nine of the Gurkhas who took part in the storming of Nilt Fort were awarded the Indian Order of Merit. In addition, Captain Aylmer, the commander of the detachment of Bengal sappers, was also awarded the Victoria Cross.

Guy Boisragon saw a great deal more active service on the North West Frontier and, in 1910, returned to his battalion, by now re-named the 5th Gurkha Rifles (Frontier Force), as its commandant. He deployed with the battalion to Egypt in 1914 and to Gallipoli in 1915.

Soldiers from the 5th Gurkha Rifles in different forms of dress circa 1900

He was shot through the kneecap in the latter action and was evacuated to the UK. He eventually reached the rank of colonel and retired from the Army in 1920. He died in France on 14 July 1931 at the age of 66.

Lieutenant J Manners-Smith VC

John (known as Jack) Manners-Smith who joined the 5th Goorkha Regiment (The Hazara Goorkha Battalion) in 1886 and who was awarded the Victoria Cross for his actions as a political officer during the Hunza-Nagar Expedition of 1891

John Manners-Smith was born in India on 30 August 1864, the son of a surgeon general in the Indian Medical Service. After serving with The Norfolk Regiment for two years, he joined the Indian Staff Corps and was appointed as a lieutenant in the 5th Goorkha Regiment (The Hazara Goorkha Battalion) on 8 February 1886. He transferred to the political department in 1887 and, in 1891, joined the Hunza-Nagar Field Force as a political officer.

'Jack' Manners-Smith (as he was known by his contemporaries) took part in the storming of the fort at Nilt after which the field force found itself unable to advance deeper into enemy territory. The problem was that the enemy had established strong-points on the high ground that dominated the approaches to its main defensive position at a place called Maiun. After 17 days of careful reconnaissance, it was eventually agreed that it might be possible to scale the vertical walls of the Thol Cliffs and seize four particular strong-points, unlocking the defensive position. The plan was that Manners-Smith and a party of 100 men from the 5th Gurkhas would scale the cliffs whilst a second group of soldiers, supported by two guns from the Hazara Mountain Battery, would fire at the four sangars to keep the enemy's heads down.

As dawn broke on 21 December 1891, Manners-Smith and his men started to climb the 1200 ft of vertical rock that led to the enemy strong-points. When they were only 60 yards from the nearest sangar, the defenders of the fort at Maiun, which was across the ravine from the sangars, saw them and began banging their drums to warn their comrades in the sangars. Alerted by the noise, the enemy left the sangars and started to throw rocks at the ascending climbers. Showing remarkable coolness, Manners-Smith and the lead Gurkhas continued to climb, eventually reaching the flatter ground on which the sangars were located. They quickly overpowered the occupants of the first sangar before moving on to the others. As Maurice Biggs notes 'the effect of this brilliant coup was to turn completely the enemy's left flank and to threaten his retreat.'[8] The field force quickly exploited this turn of events and rolled up the remainder of the enemy's positions in short order, bringing the campaign to a swift conclusion. The citation for the Victoria Cross that Manners-Smith was awarded makes exciting reading:

> For his conspicuous bravery when leading the storming party at the attack and capture of the strong position occupied by the enemy near Nilt, in the Hunza-Nagar Country, on the 20th December, 1891.
>
> The position was, owing to the nature of the country, an extremely strong one, and had

Gurkha officers of the 5th Gurkhas in uniform 1890

Gurkha soldiers of the 5th Gurkhas in native dress 1890

> barred the advance of the force for seventeen days. It was eventually forced by a small party of 50 rifles, with another of equal strength in support. The first of these parties was under the command of Lieutenant Smith, and it was entirely owing to his splendid leading, and the coolness, combined with dash, he displayed while doing so, that a success was obtained. Four nearly four hours, on the face of a cliff which was almost precipitous, he steadily moved his handful of men from point to point, as the difficulties of the ground and showers of stone from above gave him an opportunity, and during the whole of this time he was in such a position as to be unable to defend himself from any attack the enemy might choose to make.
>
> He was the first man to reach the summit, within a few yards of one of the enemy's sungars [sic], which was immediately rushed, Lieutenant Smith pistolling the first man.

Once the tribesman had surrendered and the field force had disbanded, Manners-Smith returned to his political duties. He served in numerous appointments, including as the British resident in Kathmandu from 1906–1916, and eventually finished his political service as agent to the governor general and chief commissioner of Ajmer-Merwara (Rajputana). A keen hunter, he was badly mauled by a panther in 1919. He recovered from this but died the following year from a wasting disease at the age of 55.

Captain W G Walker VC

William George Walker was born in India on 28 May 1863, the son of a deputy surgeon general. He was commissioned into the Suffolk Regiment on 29 August 1885 and joined the 4th Goorkha Regiment in 1887.

In 1903, the British despatched a force to British Somaliland to subdue Dervish tribesman led by the 'Mad Mullah', Mohammed Abdullah Hassan. The British had launched a similar but unsuccessful expedition in 1900, joining forces with the Ethiopians to try and defeat Hassan and his Dervishes. The 1903 expedition was similarly unsuccessful with Hassan's forces famously defeating the British at Gumburru and again at Daratoleh. At the time, Captain W G Walker was serving in the Somaliland Field Force with the Bikanir Camel Corps (Bikanir Ganga Risala), a unit raised by Maharaja Ganga Singh of the Indian state of Bikaner in 1889. After the defeat at Daratoleh, the British had little

William George Walker who joined the 4th Goorkha Regiment (later to become the 4th Gurkha Rifles) in May 1887 and was awarded a Victoria Cross for his actions on 22 April 1903 whilst serving with the Bikanir Camel Corps (Bikanir Ganga Risala) in Somaliland

option but to retreat. The citation for the Victoria Cross that he was subsequently awarded explains Captain Walker's part in the retreat:

During the return of Major Gough's column to Donop on the 22nd April last, after the action at Daratoleh, the rear-guard got considerably in rear of the column, owing to the thick brush, and to having to hold their ground while wounded men were being placed on camels. At this time Captain Bruce was shot through the body from a distance of about 20 yards, and fell on the path unable to move.

Captains Walker and Rolland, two men of the 2nd Battalion King's African Rifles, one Sikh and one Somali of the Camel Corps were with him when he fell.

In the meantime the column being unaware of what had happened were getting further away. Captain Rolland then ran back some 500 yards and returned with assistance to bring off Captain Bruce, while Captain Walker and the men remained with the officer, endeavouring to keep off the enemy, who were all around in the thick bush. This they succeeded in doing, though not before Captain Bruce was hit a second time, and

The adjutant and drill Instructors of the 2nd Battalion, 4th Gurkhas in 1907

British officers of the 4th Gurkhas circa 1910

> the Sikh wounded. But for the gallant conduct displayed by these Officers and men, Captain Bruce must have fallen into the hands of the enemy.

Captain Walker went on to have a distinguished military career. He commanded the 1st Battalion, 4th Gurkha Rifles, deploying with it when, in August 1914, it became the first Indian Army unit to depart India for service in the Great War. He rose quickly through the ranks and, in January 1916, was promoted to major general, assuming command of the 2nd (British) Division in France. He retired from the Army in 1919 and died on 16 February 1936 aged 72.

Rifleman Karanbahadur Rana VC

Karanbahadur Rana was born in the Baglung District of Nepal in 1898. He joined the 2nd Battalion, 3rd Queen Alexandra's Own Gurkha Rifles as a reinforcement whilst the battalion was serving in Palestine as part of 232nd Infantry Brigade in the 75th Infantry Division. As explained in Chapter 4, a number of Gurkha units were deployed to Palestine under General Sir Edmund Allenby who had taken over command of the British Forces in Egypt in June 1917. In April 1918, 2/3rd Gurkhas were engaged in an attack on a German position at a place called El Kefr. The German position, which was well sighted on the top of a rocky slope, dominated the ground to its front, putting the attacking Gurkhas at a distinct disadvantage. The citation for Rifleman Karanbahadur's Victoria Cross explains what happened:

Rifleman Karanbahadur Rana of the 2nd Battalion, 3rd Queen Alexandra's Own Gurkha Rifles who was awarded a Victoria Cross for his actions in Palestine on 10 April 1918

Maxim Gun detachment from the 2nd Battalion, 3rd Queen Alexandra's Own Gurkha Rifles circa 1918

> During an attack he, with a few other men, succeeded under intense fire, in creeping forward with a Lewis gun in order to engage an enemy machine gun which had caused severe casualties to officers and other ranks who had attempted to put it out of action. No 1 of the Lewis gun opened fire, and was shot immediately. Without a moment's hesitation Rifleman Karanbahadur pushed the dead man off the gun, and in spite of the bombs thrown at him and heavy fire from both flanks, he opened fire and knocked out the enemy machine-gun crew; then, switching his fire on to the enemy bombers and riflemen in front of him, he silenced their fire. He kept his gun in action and showed the greatest coolness in removing defects which on two occasions prevented the gun from firing. During the remainder of the day he did magnificent work, and when a withdrawal was ordered he assisted with covering fire until the enemy were close on him. He displayed throughout a very high standard of valour and devotion to duty.

King George V presented Rifleman Karanbahadur Rana with his Victoria Cross at Buckingham Palace in 1919. Karanbahadur died in Nepal on 25 July 1973 at the age of 74.

Gurkha officers of the 2nd Battalion, 3rd Queen Alexandra's Own Gurkha Rifles taken during the Great War

CHAPTER 9

Contemporary Wars: Afghanistan

2001–2015

In late December 2001, a parachute trained company of Gurkhas, known as a Gurkha Reinforcement Company (GRC), deployed to Afghanistan with the 2nd Battalion of the Parachute Regiment (2PARA). The aim of the deployment, which was known as Operation Fingal, was to assist the new Afghan Interim Authority with the provision of security in and around Kabul. From the Gurkha perspective, the deployment of C (Gurkha) Company 2PARA marked a return to operations in Afghanistan after an absence of more than 50 years. The relationship with this remarkable country continues to this day with the 2nd Battalion of The Royal Gurkha Rifles (2RGR) due to deploy to Kabul in 2015. Although both the 2001 and the next Gurkha deployments relate to

A Gurkha patrol from B Company, 1RGR on Operation Herrick 12 talking to local children in Helmand Province, Afghanistan. The Gurkhas' ability to converse with the Afghan population makes them invaluable at collecting information about local tribal dynamics

the provision of security in the country's capital, the intervening years have seen Gurkha units of all types involved in high intensity combat operations in Afghanistan's Helmand Province. This chapter draws on contemporary images to try and provide an insight into why Gurkhas have proved to be so exceptionally capable in this most challenging of operational theatres.

Although C (Gurkha) Company 2PARA returned from Afghanistan after only two months, it achieved a great deal, helping to train the first 600 strong battalion of the Afghan National Army. By the time companies from the 2nd Battalion of The Royal Gurkha Rifles deployed in Autumn 2003, the task had expanded. In addition to providing training assistance to the Afghan National Army, the battalion also had to provide a company for security duties in Kabul as well as additional troops to support the Provisional Reconstruction Teams which, following the 'hearts and minds' philosophy implemented by General Sir Gerald Templar in Malaya in 1951, were trying to improve the quality of life for rural Afghans. 2RGR's role was challenging and, inevitably, dangerous. On 12 October 2003, for example, Sergeant Kajiman Gurung was part of a six-man team tasked to go to the aid of a US convoy which had been ambushed near Kabul. Arriving at the scene, the small force began to engage the enemy. Sergeant Kajiman noticed that one of the US officers had become separated from the remainder of his eleven-man force and had sustained multiple small arms injuries. The citation for the Military Cross he was awarded for his actions explains what happened next:

> Whilst still in contact with the anti-coalition forces, Sergeant Kajiman Limbu, with little regard for his own safety, moved out from behind cover, thereby exposing himself to enemy fire from no more than fifty metres away, and dashed forward to assist the US officer. Upon reaching the injured officer, Sergeant Kajiman Limbu quickly helped him up and then supported him whilst they moved back behind cover ... Sergeant Kajiman then moved to a fire position in the centre of the friendly forces location and, using his combat weapon sight, began to engage the enemy himself whilst at the same time directing friendly forces' small arms fire onto the enemy position for an hour until the QRF[1] arrived on the scene.

Sergeant Kajiman Limbu was the first Gurkha to receive a gallantry award for operations in Afghanistan since Gurkhas ceased patrolling

Gurkhas from A Company 2RGR prepare to go on patrol

the North West Frontier in 1947. But more were to follow as the nature of British operations in Afghanistan changed dramatically only a few years later.

In 2006, the International Security Assistance Force (ISAF) began to expand its area of operations beyond Kabul. As part of this, British forces were given the task of securing Helmand Province, the largest of Afghanistan's 34 semi-autonomous provinces.

Some of the provinces, such as those in the centre and north of the country, were relatively stable when ISAF began to widen its area of operations. Others, such as those in the south and east, were much less secure. Of these, Helmand in particular was, and to some extent remains, a centre of Taliban activity. The majority of the province is

Soldiers from The Queen's Gurkha Engineers constructing a single-storey medium girder bridge in Camp Bastion, Afghanistan

open desert but the south, which borders Pakistan along the Durand Line, is mountainous. The border is hard to control and the Taliban have little difficulty bringing in both people and supplies from across the border with Pakistan.

A Gurkha sniper dominates the ground to his front on an early Herrick deployment

The vast majority of the population of Helmand, which numbers about 900,000,[2] is concentrated along the course of the River Helmand and survives through subsistence farming. The land either side of the river and its tributaries is relatively fertile and therefore tends to be cultivated. The defining features are irrigation channels (which the Gurkhas call 'nullahs'), fields and residential compounds. The latter tend to comprise a one- or two-storey house surrounded by high walls made of

A corporal from A Company 1RGR taking a well-deserved break in the cover of a cornfield during a patrol in Helmand Province, Afghanistan. 1RGR deployed on Operation Herrick 17 from October 2012 to April 2013, the battalion's second Afghan tour in 30 months

A platoon commander from A Company 1RGR leads his patrol across a river in Helmand Province, Afghanistan. A Company was taking part in Operation Kapcha Chito (Nepali for 'quick capture') and was supported by the 1st Battalion, The Duke of Lancasters (then commanded by this author's younger brother!)

mud. There is no planning authority and the villages and towns along the banks of the rivers therefore tend to be an irregular mix of nullahs, compounds, small areas of jungle and fields. To use the modern terminology, it is 'complex terrain'.

Previous page Soldiers from the 2nd Battalion, The Royal Gurkha Rifles prepare to go out on patrol in Helmand Province, Afghanistan in 2011

For the insurgent, this environment has numerous advantages, not least of which is local knowledge. But such terrain also largely negates the technological advantage that a modern army, such as the British Army, has over the majority of its adversaries. The only way to defeat the insurgent and to gain the confidence of the local population is for patrols to go into the populated areas and operate amongst the people. In this respect, there is little difference to the type of operations that the British were doing in Malaya and Borneo in the 1950s and 1960s. Soldiering in this sort of environment is both physically and mentally demanding. Patrolling in the searing heat of the Afghan summer whilst carrying body armour, ammunition, water and communications equipment requires remarkable levels of physical fitness. Add the constant threat from improvised explosive devices (IEDs) and of being ambushed by insurgents and it is easy to see why it is also requires mental resilience. Perhaps not surprisingly, Gurkhas excel in this environment. Physically

A Gurkha soldier takes a break during a foot patrol whilst his section commander consults a map. The threat from improvised explosive devices placed by the insurgents on likely patrol routes mean that soldiers often have to avoid paths and tracks

A Gurkha from B Company 1RGR adopting a fire position whilst members of his patrol talk to the local population in a village in Helmand Province, Afghanistan

and mentally robust, they also have high levels of cultural empathy with the Afghans. They understand the way the majority of Afghan people think – indeed, life in rural Afghanistan is not too dissimilar to life in the hills of Nepal.

The first Gurkha organisation to find itself operating in the new environment of Helmand Province was a composite company from across The Royal Gurkha Rifles (RGR) which deployed with 16 Air Assault Brigade on Operation Herrick 4 in April 2006. The company, known as D (Tamandu) Company, was initially tasked with guarding Camp Bastion, a new base constructed by the British in the middle of the Helmand desert.[3] However, it soon found itself conducting combat operations against the Taliban in the towns of Sangin and Now Zad, names that would become all too familiar to British soldiers and their families over the next decade. In describing the actions near Sangin to recover a downed unmanned reconnaissance aircraft, Brigadier Christopher Bullock notes that 'this successful little action represented the heaviest infantry combat the Brigade [of Gurkhas] had experienced for 40 years; there was plenty more to come.'[4]

In many ways, D (Tamandu) Company's actions in Helmand set the tone for the next nine years of combat operations by The Royal Gurkha Rifles, The Queen's Gurkha Engineers, Queen's Gurkha Signals and The Queen's Own Gurkha Logistic Regiment. Officers and soldiers from these regiments have made a profound contribution to the British Army's

Gurkhas from A Company 1RGR receiving Tika from the company's second-in-command during the battalion's Dashain Celebrations in 2012. The battalion was deployed to Afghanistan on Operation Herrick 17 from October 2012 to April 2013

many successes in Afghanistan. There have been numerous instances of individuals acting with incredible bravery. Although many of these have gone unrecorded, it is worth highlighting some of those that led to gallantry awards for the protagonists as they provide an insight into the nature of the fighting in this most recent conflict.

On 9 December 2007, a force of some 60 Taliban attacked the town of Sangin. Although they succeeded in capturing a number of Afghan National Army bases within the town, they withdrew at last light into the neighbouring villages. Determined to seize the initiative, A Company 1RGR, commanded by Major Paul Pitchfork, conducted a night infiltration of some 3.5 kilometres towards a group of villages known to be Taliban strongholds. As the company formed up at first light on 10 December 2007, it was engaged by Taliban machine guns and rocket-propelled grenades. Rifleman Bhimbahadur Gurung, the second-in-command of 3 Section of 3 Platoon, noted that a group of the enemy was about to seize a key position to the company's front. He dashed forward with his section and occupied it first. He then noticed that a second group of Taliban was using the cover of the River Helmand to try and outflank his platoon's position. The citation for the Military Cross he was awarded for his actions continues the story:

A Gurkha soldier from C Company, 1RGR adopting a fire position whilst on patrol in Afghanistan on Operation Herrick 12. 1RGR deployed with 4 Mechanised Brigade as part of Task Force Helmand from April to October 2010

A combat logistic patrol manned by soldiers from The Queen's Own Gurkha Logistic Regiment prepares to leave Camp Bastion to distribute supplies to troops deployed in patrol bases throughout Helmand Province

Bhimbahadur alone observed this counter-move. Seeing a fleeting opportunity to spoil the enemy's counter-attack, he broke cover and dashed no less than 75 metres through intense enemy fire across completely open ground to a small wall, shouting to his men to follow as he went. He then single-handedly engaged the enemy with accurate fire, halting their advance. He was subsequently joined by two machine-gunners and together they killed at least two enemy, forcing the others to withdraw. Moments later, one of the machine-gunners with Bhimbahadur Gurung was shot in the shoulder from an unseen enemy position. Bhimbahadur Gurung immediately administered first aid. He then picked up his injured colleague and, carrying him on his shoulder, extracted him back across the ground he

A young platoon commander in B Company 2RGR anoints the head of one of his soldiers in Patrol Base 'Chilly', Nad-e-Ali during Operation Herrick 14 (April to November 2011). 2RGR deployed under the command of 45 Commando for this tour

> had previously covered to the relative safety of the compound, oblivious to the continuous and withering enemy fire directed towards him.

The fighting continued for about ten hours before the enemy eventually broke contact. Major Pitchfork, the commander who sought to seize the initiative from the Taliban, was also awarded the Military Cross for '... the very highest standards of leadership, courage and initiative under fire and in the face of a most determined enemy threat.'[5]

On 17 September 2010, Acting Sergeant Dipprasad Pun was one of four men left guarding a compound after the remainder of his platoon had deployed to set up checkpoints to the east of a remote village. Sergeant Dipprasad was on duty manning a single sangar on a roof in the middle of the compound. He detected a noise and immediately suspected that his lightly manned position was about to be attacked. Having warned his commander by radio, he fired a weapon-launched grenade at the assaulting enemy and then, single-handedly, beat off attacks from multiple directions, '... killing three assailants and causing the others to flee.'[6] He was awarded the Conspicuous Gallantry Cross for his actions. As his citation notes, he '... could never have known how many enemies were attempting to overcome his position, but he sought them out from

A portrait by the artist Noreen Denzil of Acting Sergeant Dipprasad Pun of The Royal Gurkha Rifles who was awarded a Conspicuous Gallantry Cross (CGC) for his actions in defence of a remote compound in Helmand Province. To date, Acting Sergeant Dipprasad Pun is the only Gurkha soldier to have been awarded a CGC, a Gallantry Award second only to the Victoria Cross

ISAF
ISAF

Gurkhas from D Company, 1RGR engaging Taliban insurgents whilst out on patrol in Helmand Province, Afghanistan

Previous page Soldiers from The Royal Gurkha Rifles taking cover whilst out on patrol in Helmand Province, Afghanistan in 2010

A Gurkha soldier on guard duty in a sangar at FOB Delhi, one of a number of forward operating bases used by ISAF soldiers to maintain security in the villages and towns across Helmand Province

all angles despite the danger, consistently moving towards them to reach the best position of attack.'[7]

One of the most recent awards makes particularly exciting reading and, because it involves the use of the kukri, resonates with some of the Gurkhas' exploits described in the earlier chapters of this book. In March 2013, Lance Corporal Tuljung Gurung from 1RGR was on sentry duty on the front gate in a patrol base near the town of Lashkar Gah in Helmand Province. At 0345hrs, two insurgents attacked his sangar. He returned fire but was knocked to the floor when a bullet hit his helmet. As he stood up, he saw a grenade land on the floor next to him. Just as Rifleman Lachhiman Gurung had done in May 1945, he picked it up and threw it back. It exploded outside the sangar but the blast knocked him to his feet. As he started to stand up, he saw an insurgent climbing over the wall and into the sangar. Given the close range, he was unable to bring his rifle to bear so he drew his kukri, slashing at the insurgent. The citation for the Military Cross he was awarded takes up the story:

> Exposed to possible further insurgent firing positions, he aggressively and tenaciously continued to fight with his kukri. The two insurgents, defeated, turned and fled. Gurung then quickly climbed back into the sangar by which time the Quick Reaction Force had arrived. Gurung reported the incident calmly and bemoaned the fact he had not been able to prevent the insurgents escaping.[8]

Lance Corporal Tuljung Gurung from 1RGR who was awarded a Military Cross for his actions in defence of a sangar in Lashkar Gah, Helmand Province on 22 March 2013. The photograph shows Lance Corporal Tuljung at Buckingham Palace having just received his medal from His Royal Highness The Prince of Wales, Colonel-in-Chief of The Royal Gurkha Rifles and Patron of the Gurkha Welfare Trust

Gurkha soldiers from the Queen's Gurkha Signals maintaining satellite communications equipment in Camp Bastion, Helmand Province

Gurkhas have been a part of almost every operational deployment to Afghanistan since 2001. To date, officers and soldiers of the Brigade of Gurkhas have been awarded one Conspicuous Gallantry Cross, eight Military Crosses and one Distinguished Service Order in addition to many other awards and commendations. But this operational record, which is amongst the most impressive in the British Army, has come at a cost; so far, the Brigade of Gurkhas has sustained a total of 16 killed[9] in action and 51 wounded since operations began in 2001.

A soldier from The Queen's Own Gurkha Logistic Regiment conducting a house search in Helmand Province

Although Britain has now ceased combat operations in Afghanistan, it continues to commit forces to NATO's Resolute Support (RS) Mission. This aims to support the Afghan National Security Forces as they take the lead for the country's security. It is an important mission and Gurkhas are playing a key role in its delivery. But it is not without risk and that is why continued support for the work of the Gurkha Welfare Trust is vital. As His Royal Highness The Prince of Wales outlines in his introduction, the Trust 'exists to enable retired Gurkhas to live out their lives with dignity; providing welfare services – ranging from pensions and residential homes for ex-Gurkhas, through to schools and water projects in the most remote regions of Nepal.' With Britain's Gurkhas still in the thick of the action, there is little doubt that the Trust's services will be required for many years to come.

A Gurkha officer salutes the Memorial to the Fallen in the British Gurkhas Nepal (BGN) Camp in Kathmandu during a Garrison Remembrance Parade. To date, 16 members of the Brigade of Gurkhas have been killed in Afghanistan and a further 51 wounded since the first Gurkha troops deployed in 2001. The memory of those who sacrificed their lives in this campaign, along with those who died fighting for the Crown in previous conflicts, lives on in the brigade. They will never be forgotten

Appendix 1

Battle Honours of the Brigade of Gurkhas

The battle honours of those regiments that remained in the Indian Army after 1947 are only shown up until that date.

1st King George V's Own Gurkha Rifles (The Malaun Regiment)

Bharatpur, Aliwal, Sobraon, Afghanistan 1878–80, Tirah, Punjab Frontier, Givenchy 1914, Neuve Chapelle, Ypres 1915, St. Julien, Festubert 1915, Loos, France and Flanders 1914–15, Megiddo, Sharon, Palestine 1918, Tigris 1916, Kut-al-Amara 1917, Baghdad, Mesopotamia 1916–18, North-West Frontier, India 1915, 1917, Afghanistan 1919, Jitra, Kampar, Malaya 1941–42, Shenam Pass, Bishenpur, Ukhrul, Myinmu Bridgehead, Kyaukse 1945, Burma 1942–45

2nd King Edward VII's Own Gurkhas (The Sirmoor Rifles)

Bhurtpore, Aliwal, Sobraon, Delhi 1857, Kabul 1879, Kandahar 1880, Afghanistan 1878–80, Tirah, Punjab Frontier, La Bassée 1914, Festubert 1914–15, Givenchy 1914, Neuve Chapelle, Aubers, Loos, France and Flanders 1914–15, Egypt 1915, Tigris 1916, Kut-al-Amara 1917, Baghdad, Mesopotamia 1916, 1918, Persia 1918, Baluchistan 1918, Afghanistan 1919, El Alamein, Mareth, Akarit, Djebel el Meida, Enfidaville, Tunis, North Africa 1942–43, Cassino I, Monastery Hill, Pian di Maggio, Gothic Line, Coriano, Poggio San Giovanni, Monte Reggiano, Italy 1944–45, Greece 1944–45, North Malaya, Jitra, Central Malaya, Kampar, Slim River, Johore, Singapore Island, Malaya 1941–42, North Arakan, Irrawaddy, Magwe, Sittang 1945, Point 1433, Arakan Beaches, Myebon, Tamandu, Chindits 1943, Burma 1943–45

3rd Queen Alexandra's Own Gurkha Rifles

Delhi 1857, Ahmed Khel, Afghanistan 1878–80, Burma 1885–87, Chitral, Tirah, Punjab Frontier, La Bassée 1914, Armentieres 1914, Festubert 1914, 1915, Givenchy 1914, Neuve Chapelle, Aubers, Loos, France and Flanders 1914–15, Egypt 1915–1916, Gaza, El Mughar, Nebi Samwil, Jerusalem, Tell, Asur, Megiddo, Sharon, Palestine 1917–18, Sharqat, Mesopotamia 1917–18, Afghanistan 1919, Deir el Shein, North Africa 1940–43, Monte della Gorgace, Il Castello, Monte Farneto, Monte Cavallo, Italy 1943–45, Sittang 1942, Kyaukse 1942, Imphal, Tuitum, Sakawng, Shenam Pass, Bishenpur, Tengnoupal, Meiktila, Defence of Meiktila, Rangoon Road, Pyawbwe, Pegu 1945, Burma 1942–45

4th Prince of Wales's Own Gurkha Rifles

Ali Masjid, Kabul 1879, Kandahar 1880, Afghanistan 1878–80, Waziristan 1895, Chitral, Tirah, Punjab Frontier, China 1900, Givenchy 1914, Neuve Chapelle, Ypres 1915, St. Julien, Aubers, Festubert 1915, France and Flanders 1914–15, Gallipoli 1915, Egypt 1916, Tigris 1916, Ku-al-Amara 1917, Baghdad, Mesopotamia 1916–18, North-West Frontier, India 1917, Baluchistan 1918, Afghanistan 1919, Iraq 1941, Syria 1941, The Cauldron, North Africa 1940–43, Trestina, Monte Cedrone, Italy 1943–45, Pegu 1942, Chindits 1944, Mandalay, Burma 1942–45, Bishenpur, Shwebo

5th Royal Gurkha Rifles (Frontier Force)

Peiwar Kotal, Charasiah, Kabul 1879, Kandahar 1880, Afghanistan 1878–80, Punjab Frontier, Helles, Krithia, Suvla, Sari Bair, Gallipoli 1915, Suez Canal, Egypt 1915–16, Khan Baghdadi, Mesopotamia 1916–18, North-West Frontier, India 1917, Afghanistan 1919, The Sangro, Caldari, Cassino II, Sant'Angelo in Teodice, Rocca d'Arce, Ripa Ridge, Femmina Morte, Monte San Bartolo, The Senio, Italy 1943–45, Sittang 1942, 1945, Kyaukse 1942, Yenangyaung 1942, Stockades, Buthidaung, Imphal, Sakawng, Bishenpur, Shenam Pass, The Irrawaddy, Burma 1942–45

6th Queen Elizabeth's Own Gurkha Rifles

Burma 1885–87, Helles, Krithia, Suvla, Sari Bair, Gallipoli 1915, Suez Canal, Egypt 1915–16, Khan Baghdadi, Mesopotamia 1916–18, Persia 1918, North-West Frontier, India 1915, Afghanistan 1919, Coriano, Santacangelo, Monte Chicco, Lamone Crossing, Senio Floodbank, Medicina, Gaiana Crossing, Italy 1944–45, Shwebo, Kyaukmyaung Bridgehead, Mandalay, Fort Dufferin, Maymyo, Rangoon Road, Toungoo, Sittang 1945, Chindits 1944, Burma 1942–45

7th Duke of Edinburgh's Own Gurkha Rifles

Suez Canal, Egypt 1915–16, Megiddo, Sharon, Palestine 1918, Shaiba, Kut-al-Amara 1915, 1917, Ctesiphon, Defence of Kut-al-Amara, Baghdad, Sharqat, Mesopotamia 1915–1918, Afghanistan 1919, Tobruk 1942, North Africa 1942, Cassino I, Campriano, Poggio del Grillo, Tavoleto, Montebello Scorticata Ridge, Italy 1944, Sittang 1942, 1945, Pegu 1942, Kyaukse 1942, Shwegyin, Imphal, Bishenpur, Meiktila, Capture of Meiktila, Defence of Meiktila, Rangoon Road, Pyawbwe, Burma 1942–45, Falkland Islands 1982

8th Gurkha Rifles

Burma 1885–87, La Bassée 1914, Festubert 1914, 1915, Givenchy 1914, Neuve Chapelle, Aubers, France and Flanders 1914–15, Egypt 1915–16, Megiddo, Sharon, Palestine 1918, Tigris 1916, Kut-al-Amara 1917, Baghdad, Mesopotamia 1916–17, Afghanistan 1919, Iraq 1941, North Africa 1940–43, Gothic Line, Italy 1943–45, Coriano, Santarcangelo, Gaiana Crossing, Point 551, Imphal, Tamu Road, Bishenpur, Kanglatongbi, Mandalay, Myinmu Bridgehead, Singhu, Shandatgyi, Sittang 1945, Burma 1942–45

9th Gurkha Rifles

Bhurtpore, Sobraon, Afghanistan 1879–80, Punjab Frontier, La Bassée 1914, Armentieres 1914, Festubert 1914, 1915, Givenchy 1914, Neuve Chapelle, Aubers, Loos, France and Flanders 1914–15, Tigris 1916, Kut-al-Amara 1917, Baghdad, Mesopotamia 1916–18, Afghanistan 1919, Djebel el Meida, Djebel Garci, Ragoubet Souissi, North Africa 1940–43, Cassino I, Hangman's Hill, Tavoleto, San Marino, Italy 1943–45, Greece 1944–45, Malaya 1941–42, Chindits 1944, Burma 1942–45

10th Princess Mary's Own Gurkha Rifles

Amboor, Carnatic, Mysore, Assaye, Ava, Burma 1885–87, Helles, Krithia, Suvla, Sari Bair, Gallipoli 1915, Suez Canal, Egypt 1915, Sharqat, Mesopotamia 1916–18, Afghanistan 1919, Iraq 1941, Deir es Zor, Syria 1941, Coriano, Santarcangelo, Senio Floodbank, Bologna, Sillaro Crossing, Gaiana Crossing, Italy 1944–45, Monywa 1942, Imphal, Tuitum, Tamu Road, Shenam Pass, Litan, Bishenpur, Tengnoupal, Mandalay, Myinmu Bridgehead, Kyaukse 1945, Meiktila, Capture of Meiktila, Defence of Meiktila, Irrawaddy, Rangoon Road, Pegu 1945, Sittang 1945, Burma 1942–45

11th Gurkha Rifles

Afghanistan 1919

The Royal Gurkha Rifles

Amboor, Carnatic, Mysore, Assaye, Ava, Bhurtpore, Aliwal, Sobraon, Delhi 1857, Kabul 1879, Kandahar 1880, Afghanistan 1878–80, Burma 1885–87, Tirah, Punjab Frontier, La Bassée 1914, Festubert 1914, 1915, Givenchy 1914, Neuve Chappelle, Aubers, Loos, France and Flanders 1914–15, Helles, Krithia, Suvla, Sari Blair, Gallipoli 1915, Suez Canal, Megiddo, Egypt 1915–16, Sharon, Palestine 1918, Shaiba, Kut-al-Amara 1915, 1917, Ctesiphon, Defence of Kut-al-Amara, Tigris 1916, Baghdad, Khan Baghdadi, Sharqat, Mesopotamia 1915–1918, Persia 1918, North West Frontier, India 1915, Baluchistan 1918, Afghanistan 1919, Iraq 1941, Deir es Zor, Syria 1941, Tobruk 1942, El Alamein, Mareth, Akarit, Djebel el Meida, Enfidaville, Tunis, North Africa 1942–43, Cassino 1, Monastery Hill, Pian de Maggio, Campriano, Poggio Del Grillo, Gothic Line, Tavoleto, Coriano, Poggio San Giovanni, Montebello-Scorticata Ridge, Santacangelo, Monte Reggiano, Monte Chicco, Lamone Crossing, Senio Floodbank, Bologna, Sillaro Crossing, Medicina, Gaiana Crossing, Italy 1944–45, Greece 1944–45, North Malaya, Jitra, Central Malaya, Kampar, Slim River, Johore, Singapore Island, Malaya 1941–42, Sittang 1942, 1945, Pegu 1942, 1945, Kyaukse 1942, 1945, Monywa 1942, Shwegyin, North Arakan, Imphal, Tuitum, Tamu Road, Shenam Pass, Litan, Bishenpur, Tengnoupal, Shwebo, Kyaukmyaung Bridgehead, Mandalay, Myinmu Bridgehead, Fort Dufferin, Maymo, Meiktila, Capture of Meiktila, Defence of Meiktila, Irrawaddy, Magwe, Rangoon Road, Pyabwe, Toungoo, Point 1433, Arakan Beaches, Myebon, Tamandu, Chindits 1943, 1945, Burma 1942–45, Falkland Islands 1982

Appendix 2

The Current Brigade of Gurkhas

The British Army is currently reducing down to its new size of around 82,000 Regular personnel. Within this, the strength of the Brigade of Gurkhas will be 2617 personnel by 2020.

The vast majority of Gurkhas serve in formed units, companies or troops within the Brigade of Gurkhas. Some Gurkhas are also serving in specialist units (such as the Special Air Service) having passed the rigorous selection tests associated with these organisations.

The Royal Gurkha Rifles (RGR)

The RGR is organised into two infantry battalions (one in the UK and one in Brunei). Small detachments of RGR personnel also exist at the Royal Military Academy Sandhurst, the Infantry Battle School in Brecon and the Infantry Training Centre in Catterick.

The RGR's strength is 1261, accounting for some 49 per cent of the Brigade of Gurkhas.

The RGR accounts for 6 per cent of the Regular Infantry strength in the British Army.

The Queen's Gurkha Engineers (QGE)

The QGE is comprised of a Regimental HQ (located within 36 Engineer Regiment) and two squadrons (69 and 70 Gurkha Field Squadrons).

The QGE strength is 295, making up 11 per cent of the Brigade of Gurkhas.

The QGE accounts for 4 per cent of Royal Engineer manpower.

The Queen's Own Gurkha Logistic Regiment (QOGLR)

The QOGLR is organised into a single regiment, 10 QOGLR, which is comprised of three squadrons (1 Squadron, 28 Squadron and 36 HQ Squadron).

The QOGLR's strength is 439, accounting for 16 per cent of the Brigade of Gurkhas.

The QOGLR accounts for 4 per cent of Royal Logistic Corps manpower.

Queen's Gurkha Signals (QGS)

QGS is comprised of three squadrons (246, 248 and 250 Squadrons) and a number of smaller troops (based in the Royal Military Academy Sandhurst, Nepal and Brunei).

QGS's strength is 484, accounting for 19 per cent of the Brigade of Gurkhas.

QGS accounts for 7 per cent of Royal Signals manpower.

The Band of the Brigade of Gurkhas

The Band of the Brigade of Gurkhas has strength of 45, accounting for 2 per cent of the Brigade of Gurkhas.

The Band accounts for 6 per cent of the manpower in the Corps of Army Music.

Gurkha Staff and Personnel Support (GSPS)

There are 93 GSPS personnel deployed across the Brigade of Gurkhas and the wider British Army. They account for 4 per cent of the Brigade of Gurkhas.

GSPS personnel make up 3 per cent of the Adjutant General's Corps (AGC).

Appendix 3

The Gurkha Welfare Trust

Gurkhas young and old. The Gurkha Welfare Trust helps ex-Gurkhas live out their lives with dignity

The Gurkha Welfare Trust is the leading Gurkha welfare charity. It was established in 1969 when it was realised that a significant number of Gurkha veterans were living in their homeland of Nepal in circumstances of abject poverty and distress. These were predominantly soldiers who had been discharged at the end of the Second World War without entitlement to an Army pension.

An appeal, led by the British officers of the Brigade of Gurkhas, was made to the British public. Incredibly, over £1 million was raised, truly indicative of the esteem in which the Gurkha soldier was held. The Gurkha Welfare Trust was born.

Its mission was, and remains to this day, to ensure that no Gurkha veteran or his dependants live in distress. In the succeeding 45 years, its work has developed to meet changing needs.

Today, the Trust provides a monthly **Welfare Pension** to 6667 ex-servicemen and widows. Currently set at 8000 Nepalese rupees (around £53), the pension is a vital financial lifeline and is often the only source of income for a family.

The Trust is also there to support Gurkha veterans through difficult times – by looking after their medical needs through its **Medical Programme** and by providing one-off emergency **Hardship Grants** in times of crises.

A Gurkha veteran and Trust Welfare Pensioner outside his home

Rifleman Kajibahadur Limbu, a resident at The Rambahadur Limbu VC Home in Dharan, east Nepal

For the frailest, the Trust provides 24-hour care in two **Residential Homes** in Pokhara and Dharan.

It also pays an annual **Winter Fuel Allowance** to cover the additional costs incurred during the colder winter months.

The Trust also recognises that it has a duty to the hill communities from which Gurkha soldiers are recruited and to which they, until recently, returned after service. Nepal is a land of extraordinary beauty and variety. There is another Nepal, however. Rugged, harsh, unforgiving and sparsely populated, life for many is one of

The Kulbir Thapa VC Home in Pokhara, west Nepal

A Nepali villager on her way to fetch water in Pani Pokhara, a remote village in northern Nepal

A tap stand constructed by the Gurkha Welfare Trust

A girl drinking from a tap stand in Bharat Pokhari, a village near Pokhara, following a GWT project in 2013

A plaque at Kalika Secondary School in Myagdi district, built by the Gurkha Welfare Trust in 2011

Balkalya Higher Secondary School in Libang, west Nepal, constructed by the GWT in 2002

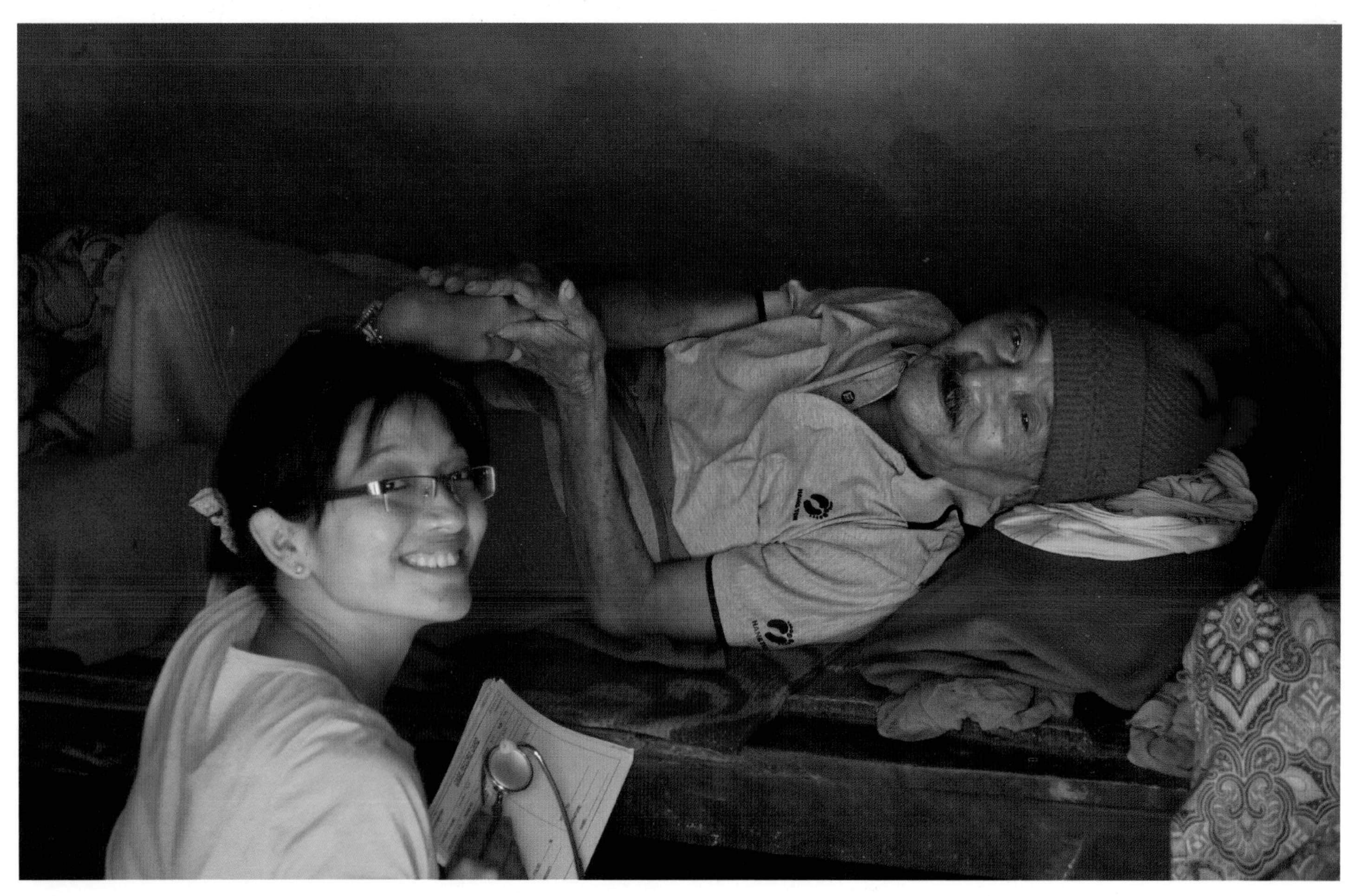

Mobile Doctor Sobimaya Tamang with Rifleman Ganghar Bhar Thapa during a home visit

subsistence where extremes of climate can destroy lives and livelihoods and where medical and other support can be several days' walk away.

To date, its **Rural Water and Sanitation Programme** has installed water and sanitation schemes in 1406 hill villages, bringing the gift of clean water to 297,792 people. Its **School Programme** has constructed 127 schools and extended, repaired or refurbished a further 1544 schools, benefitting 546,504 local children. Each year it runs eight **Mobile Medical Camps**, providing basic medical, dental, gynaecological and ophthalmic care to over 14,500 people.

In the UK, it runs the only specialist advice and information service for ex-Gurkha servicemen and their families through two **Welfare Advice Centres** in Salisbury and Aldershot.

The Trust needs to raise around £13 million each year to fund this vital work. The funds raised through the sale of this book will be used to enhance its Medical Programme; to provide more mobile doctors, more district nurses and greater support for families caring for elderly Gurkha veterans in their final years. Thank you for your support!

The Gurkha Welfare Trust, PO Box 2170, 22 Queen Street, Salisbury SP2 2EX
www.gwt.org.uk Tel: 01722 323955 Email: info@gwt.org.uk

Appendix 4

The Gurkha Museum

The fascinating story of Gurkha service to the Crown since 1815 is presented at The Gurkha Museum in Winchester. Gurkha participation in the epic periods of world military history over the last 200 years is portrayed in graphic detail using artefacts, text, dioramas, tableaux and exciting touch screen technology. The museum records the 26 Victoria Crosses and many other medals awarded to Gurkha regiments. Gurkhas have seen action all around the world, particularly during the two world wars when their bravery and reputation as exceptional soldiers was recognised with the award of twelve of these Victoria Crosses. Such bravery extended into the jungles of Borneo in 1965 with the award of a further Victoria Cross and to the deserts of Afghanistan in 2010 with Sergeant Dipprasad Pun of The Royal Gurkha Rifles being awarded the Conspicuous Gallantry Cross.

A Gurkha Museum was first established at Queen Elizabeth Barracks at Church Crookham in Hampshire from items collected from the serving units of the Brigade of Gurkhas and other donations. It was opened on 21 June 1974 by Field Marshal The Lord Harding of Petherton. From the start the museum was set up to represent all the units of the brigade. As the museum, housed in a wooden barrack block, grew and developed it was clear that larger and more permanent premises were needed. In April 1989

The old Rifle Depot at Peninsula Barracks in Winchester, home of the Gurkha Museum since 1990

Captain Rambahadur Limbu VC last visited the Gurkha Museum in 2014. He is shown here looking at a display which explains the remarkable actions for which he was awarded a VC as a lance corporal during the Borneo Confrontation. The story, including an extract from his citation, is covered in detail in Chapter 8

the museum at Church Crookham was closed and artefacts began being transferred to a fully refurbished building at the old Rifle Depot at Peninsula Barracks in Winchester. Field Marshal The Lord Bramall of Bushfield officially opened the new Museum on 16 July 1990.

Since then the museum has expanded its collection and archive. Today it records the service of six current capbadged units of the brigade as well as the many Gurkha regiments and units which existed prior to 1994. The archive contains an unrivalled collection of books, documents, photographs and film, not just recording Gurkha military history but also the country and people of Nepal. Indeed the cultural aspects of service in Britain's Brigade of Gurkhas form an important part of the displays. The museum is the focal point for the serving brigade's corporate memory and heritage and strives to develop a strong sense of connection with its past, present and future. As the brigade moves on from operations in Afghanistan, the museum will continue to record its history and role in the British Army, here in the UK, in Brunei and wherever it is deployed in the world.

The Gurkha Museum welcomes visits, particularly in this 200th year of Gurkha service to the Crown, from members of the public who wish to learn more about the fascinating history of Britain's Gurkhas.

Notes

CHAPTER 1

1 Thorn (1816), p239.
2 Pemble (2009), p369.
3 Prinsep (1820), p46.
4 Caplan (1995), p14.
5 Bullock (2009), p22.
6 Parker (2000), p40.
7 Parker (2000), p43.
8 Bullock (2009), p22.
9 Caplan (1995), p11.
10 Prinsep (1820), p47.
11 Prinsep (1820), p47.
12 Prinsep (1820), p47.
13 Prinsep (1820), p47.
14 Pemble (2009), p370.
15 Wilson (1858), p16.
16 Pemble (2009), p370.
17 Prinsep (1820), p46.
18 Prinsep (1820), p105.
19 Pemble (2009), p362.
20 Parker (2000), p44.
21 Prinsep (1820), pp104–105.
22 Prinsep (1820), p107.
23 Prinsep (1820), p108.
24 Wilson (1858), p49.
25 Prinsep (1820), p109.
26 Prinsep (1820), p109.
27 Wilson (1858), p50.
28 Prinsep (1820), p109.
29 Bullock (2009), p27.
30 Bullock (2009), p27.
31 Parker (2000), p42.
32 Prinsep (1820), p110.
33 Bullock (2009), p27.
34 Bullock (2009), p26.
35 Bullock (2009), p26.
36 Parker (2000), p46.
37 Gould (1999), p59.

CHAPTER 2

1 Bullock (2009), p31.
2 Shore (1836), pp44–45.
3 Shore (1836), p41.
4 Shore (1836), p45.
5 Shore (1836), p45.
6 Shore (1836), p45.
7 Shore (1836), p45.
8 Shore (1836), p45.
9 Bullock (2009), p29.
10 Hill (1858), p128.
11 Hill (1858), p128.
12 Hill (1858), p128.
13 Hill (1858), p128.
14 Hill (1858), p129.
15 Hill (1858), p134.
16 Bullock (2009), p32.
17 Marshall (2006), p548.
18 Bullock (2009), p32.
19 Marshall (2006), p548.
20 Hill (1858), p138.
21 Hill (1858), p139.
22 Hill (1858), p141.
23 Hill (1858), p142.
24 Woodburn (2012), p146.
25 Woodburn (2012), p146.
26 Bullock (2009), p33.
27 Bullock (2009), p33.
28 Bullock (2009), p33.
29 Woodburn (2012), p146.
30 Woodburn (2012), p146.
31 Encyclopedia Britannica Online.
32 Bullock (2009), p33.
33 Bullock (2009), p33
34 Bullock (2009), p33.
35 Parker (2000), p56.
36 Bullock (2009), p35.
37 Bullock (2009), p36.
38 Parker (2000), p58.
39 Parker (2000), p59.
40 Bullock (2009), p39.
41 Parker (2000), p38.
42 Bullock (2009), p38.
43 Parker (2000), p59.
44 Bullock (2009), 39.
45 Parker (2000), p60.
46 Parker (2000), p60.
47 Read and Fisher (1998), p57.
48 Parker (2000), p61.
49 Biggs (2012), p21.
50 Bullock (2009), p39.
51 Bullock (2009), p39.
52 Parker (2000), p62.
53 Parker (2000), p62.
54 Parker (2000), p62.
55 Read and Fisher (1998), p55
56 Read and Fisher (1998), p55.
57 Parker (2000), p63.

CHAPTER 3

1 Morris (1979), p91.
2 Parker (2000), p76.
3 Morris (1979), pp 90–91.
4 Bullock (2009), p47.
5 Morris (1979), p92.
6 Morris (1979), p92.
7 Morris (1979), p94.
8 Morris (1979), p93.
9 Bullock (2009), p47.
10 Morris (1979), p95.
11 Morris (1979), p95.
12 Bullock (2009), p49.
13 Bullock (2009), p49.
14 Morris (1979), p105.
15 Morris (1979), p107.
16 Morris (1979), p102.
17 Morris (1979), p108.
18 Morris (1979), p112.
19 Morris (1979), p418.
20 Bullock (2009), p50.
21 Parker (2000), p77.
22 Bullock (2009), p51.
23 Parker (2000), p77.
24 Parker (2000), p77.
25 Parker (2000), p76-77.
26 Parker (2000), p78.
27 Bullock (2009), p51.
28 Bullock (2009), p51.
29 Bullock (2009), p51.
30 Bullock (2009), p51.
31 Bullock (2009), p51.
32 Parker (2000), p78.
33 Bullock (2009), p51.
34 Bullock (2009), p52.
35 Bullock (2009), p52.
36 Bullock (2009), p51.
37 Bullock (2009), p52.
38 Bullock (2009), p52.
39 Bullock (2009), p53.
40 Bullock (2009), p53.
41 Bullock (2009), p53.
42 Parker (2000), p76.
43 Raugh (2004), p163.
44 Raugh (2004), p163.
45 Raugh (2004), p164.
46 Raugh (2004), p164.
47 Parker (2000), p82.
48 Parker (2000), p82.
49 Bullock (2000), p54.
50 Parker (2000), p82.
51 Bullock (2000), p54.
52 Bullock (2000), p54.
53 Bullock (2000), p55.
54 Bullock (2009), p55.
55 Parker (2000), p84.
56 Parker (2000), p85.
57 Parker (2000), p85.
58 Carey and Tuck (1896), p15.
59 Parker (2000), p85.
60 Carey and Tuck (1896), p16.
61 Low (1883), p111.
62 Parker (2000), p85.
63 Parker (2000), p85.
64 Parker (2000), p87.
65 Bullock (2009), p44.
66 Bullock (2009), p44.
67 Parker (2000), p86.
68 London Gazette (1872), p4489.
69 Parker (2000), p86.
70 Carey and Tuck (1896), p16.
71 Bullock (2009), p44.
72 Parker (2000), p88
73 Parker (2000), p88.
74 Bullock (2009), p45.
75 Bullock (2009), p45.
76 Parker (2000), p89.
77 Bullock (2000), p45.
78 New Zealand Herald (1891), p6.
79 Bullock (2009), p45.
80 Parker (2009), p89.
81 The Argus (1891), p5.
82 Parker (2000), p90.
83 Parker (2000), p90.
84 Bullock (2009), p45.
85 Bullock (2009), p45.
86 Parker (2000), p91.
87 Bullock (2000), p45.
88 Parker (2000), p92.
89 New Zealand Herald (1891), p6.
90 Palace (2004), p6.
91 Palace (2004), p2.
92 Palace (2004), p5.
93 Parker (2000), p92.
94 Parker (2000), p93.
95 Parker (2000), p93.
96 Parker (2000), p93.
97 Palace (2004), p7.
98 Palace (2004), p7.

CHAPTER 4

1 Bullock (2009), p58.
2 Parker (2000), p101.
3 Bullock (2009), p63.
4 Bullock (2009), p56.
5 Bullock (2009), p57.
6 Bullock (2009), p57.
7 Bullock (2009), p57.
8 Parker (2000), p101.
9 Parker (2000), p101.
10 Bullock (2009), p60.
11 Parker (2000), p102.
12 Parker (2000), p102.
13 Parker (2000), p60.
14 Bullock (2000), p60.
15 Bullock (2009), p63.
16 Bullock (2009), p64.
17 Trench (1988), p41.
18 Bullock (2009), p64.
19 Trench (1988), p41.
20 Bullock (2009), p65.
21 Parker (2000), p103.
22 Parker (2000), p104.
23 Parker (2000), p109.
24 Parker (2000), p110.
25 Parker (2000), p110.
26 Parker (2000), p111.
27 London Gazette (1915), p11, 450.
28 Parker (2000), p112.
29 Trench (1988), p44.
30 Trench (1988), p44.
31 Trench (1988), p44.
32 Parker (2000), p113.
33 Bullock (2009), p65.
34 Ross (2005), p18.
35 Ross (2005), p18.
36 Ross (2005), p18.

37 Parker (2000), p116.
38 Parker (2000), p118.
39 Parker (2000), p118.
40 Bullock (2009), p68.
41 Parker (2000), p118.
42 Parker (2000), p118.
43 Bullock (2009), p69.
44 Parker (2000), p118.
45 Parker (2000), p119.
46 Parker (2000), p119.
47 Parker (2000), p121.
48 Bullock (2009), p72.
49 Parker (2000), p122.
50 Parker (2000), p123.
51 Parker (2000), p123.
52 Lunt (1994), p144
53 "August 2", http://www.gallipoli-association.org/on-this-day/august-2 accessed on 16 October 2014.
54 "August 2", http://www.gallipoli-association.org/on-this-day/august-2 accessed on 16 October 2014.
55 Ross (2005), p20.
56 Bullock (2009), p72.
57 Parker (2000), p124.
58 Parker (2000), p126.
59 Marshall Cavendish (2002), p196.
60 Bullock (2009), pp 73–74.
61 Murland (2012), p200.
62 Bullock (2009), p74.
63 Murland (2012), p200.
64 Parker (2000), p128.
65 Marshall Cavendish (2002), p197.
66 Bullock (2009), p74.
67 Murland (2012), p200.
68 Bullock (2009), p74.
69 Murland (2012), p200.
70 Parker (2000), p129.
71 Murland (2012), p200.
72 Murland (2012), p200.
73 Marshall Cavendish (2002), p198.
74 Bullock (2009), p74.
75 Parker (2000), p129.
76 Bullock (2009), p74.
77 Murland (2012), p201.
78 Bullock (2009), p74.
79 Parker (2000), p129.
80 Marshall Cavendish (2002), p198.
81 Parker (2000), p130.
82 Meyers (2004).
83 Parker (2000), p131.
84 Parker (2000), p131.
85 Marshall Cavendish (2002), p199.
86 Parker (2000), p132.
87 Bullock (2000), p77.
88 Bullock (2009), p78.
89 Bellamy (2011), pXX
90 Bullock (2009), p79.
91 Bullock (2000), p80.
92 Knight (2013), p143.
93 Knight (2013), p143.
94 Bullock (2009), p82.
95 Bullock (2009), p82.
96 Knight (2013), p143.
97 Bullock (2009), p83.
98 Mortlock (2010), p101.
99 Bullock (2009), p82.
100 Murphy (2012), p33.
101 Marshall Cavendish (2002b), p492.
102 Marshall Cavendish (2002b), p492.
103 Marshall Cavendish (2002b), p492.

CHAPTER 5

1 Ferguson (2003), p330.
2 Ferguson (2003), p331.
3 Read and Fisher (1998), p168.
4 Read and Fisher (1998), p169.
5 Read and Fisher (1998), p169.
6 Read and Fisher (1998), pp 169–170.
7 Read and Fisher (1998), p171.
8 Ferguson (2003), p332.
9 Ferguson (2003), p333.
10 Ferguson (2003), p332.
11 Bullock (2009), pp 85–86.
12 Ferguson (2003), p332.
13 Ferguson (2003), p332.
14 Ferguson (2003), p333.
15 Read and Fisher (1998), p171.
16 Collett (2005), p424.
17 Johnson (2011), p180.
18 Johnson (2011), p180.
19 Johnson (2011), p180.
20 Bullock (2009), p87.
21 Bullock (2011), p87.
22 Johnson (2011), p183.
23 Johnson (2011), p183.
24 Bullock (2011), p87.
25 Johnson (2011), p184.
26 Bullock (2011), p87.
27 Johnson (2011), p180.
28 Johnson (2011), p185.
29 Johnson (2011), p185.
30 Johnson (2011), p185.
31 Bullock (2009), p88.
32 Johnson (2011), p185.
33 Johnson (2011), p185.
34 Bullock (2009), p185.
35 Johnson (2011), p185.
36 Johnson (2011), p186.
37 Johnson (2011), p186.
38 Bullock (2009), p87.
39 Johnson (2011), p187.
40 Johnson (2011), p187.
41 Johnson (2011), p187.
42 Johnson (2011), p187.
43 Johnson (2011), p187.
44 Johnson (2011), p187.
45 Johnson (2011), p187.
46 Johnson (2011), p187.
47 Johnson (2011), p187.
48 Johnson (2011), p187.
49 Johnson (2011), p188.
50 Bullock (2009), p88.
51 Johnson (2011), p188.
52 Trench (1988), p104.
53 Trench (1988), p104.
54 Trench (1988), p104.
55 Trench (1988), p104.
56 Bullock (2009), p89.
57 Trench (1988), p104.
58 Trench (1988), p104.
59 Trench (1988), p109.
60 Trench (1988), p105.
61 Trench (1988), p105.
62 Trench (1988), p106.
63 Trench (1988), p106.
64 Trench (1988), p106.
65 Trench (1988), p107.
66 Trench (1988), p109.
67 Trench (1988), p110.
68 Trench (1988), p110.
69 Trench (1988), p110.
70 Bullock (2009), p89.
71 Trench (1988), p111.
72 Trench (1988), p111.
73 Trench (1988), p111.
74 Trench (1988), p112.
75 Trench (1988), p195.
76 Trench (1988), p195.
77 Trench (1988), p196.
78 Bullock (2009), p93.
79 Trench (1988), p197.
80 Bullock (2009), p93.
81 Trench (1988), p198.
82 Trench (1988), p198.
83 Trench (1988), p198.

CHAPTER 6

1 Trench (1988), p137
2 Bullock (2009), p97.
3 Bullock (2009), p98.
4 Bullock (2009), p100.
5 Bullock (2009), p100.
6 Bullock (2009), pp 100–101.
7 Bullock (2009), p101.
8 Bullock (2009), p101.
9 Bullock (2009), p101.
10 Bullock (2009), p102.
11 Bullock (2009), p102.
12 Bullock (2009), p102.
13 Bullock (2009), p102.
14 Trench (1988), p224.
15 Bullock (2009), p108.
16 Trench (1988), p225.
17 Bullock (2009), p108.
18 Trench (1988), p170.
19 Trench (1988), p225.
20 Trench (1988), p225.
21 Trench (1988), p225.
22 Bullock (2009), p108.
23 Trench (1988), p226.
24 Bullock (2009), p108.
25 Biggs (2012), p50.
26 Trench (1988), p226.
27 Bullock (2009), p110.
28 Biggs (2012), p50.
29 Baudot and Dilson (1981), p320.
30 Baudot and Dilson (1981), p319.
31 Bullock (2009), p112.
32 Bullock (2009), p115.
33 Bullock (2009), p116.
34 Bullock (2009), p117.
35 Bullock (2009), p118.
36 Bullock (2009), p118.
37 Bullock (2009), p118.
38 http://www.victoriacross.co.uk/descrip s.html accessed on 27 December 2014.
39 Biggs (2012), p68.
40 Biggs (2012), p68.
41 Bullock (2009), p121.
42 Biggs (2012), p70.
43 Bullock (2009), p123.
44 Bullock (2009), p124.
45 Gibbs (1955), p103.
46 Baudot and Dilson (1981), p385.
47 Baudot and Dilson (1981), p385.
48 Bullock (2009), p127.
49 Walker (1997), p84.
50 Bullock (2009), p136.
51 Bullock (2009), p139.
52 Trench (1997), p205.
53 Trench (1997), p203.
54 Bullock (2009), p140.
55 Walker (1997), p63.
56 Walker (1997), p63.
57 Bullock (2009), p145.
58 Bullock (2009), p146.
59 Bullock (2009), p145.
60 Bullock (2009), p146.
61 Bullock (2009), p147.
62 Trench (1988), p255.
63 Trench (1988), p255.
64 Biggs (2012), p52.
65 Biggs (2012), p52.
66 Biggs (2012), p52.
67 Biggs (2012), p53. The Gurkha battle-cry was (and remains) 'Ayo Gurkhali' (the Gurkhas are upon you!).
68 Bullock (2009), p147.
69 Bullock (2009), p149.
70 Bullock (2009), p154.
71 Bullock (2009), p154.
72 Bullock (2009), pp 154–155.
73 Bullock (2009), p157.
74 Chapple (2014), p185
75 Biggs (2012), p56.
76 Biggs (2012), p57.
77 Biggs (2012), p60.
78 Biggs (2012), p61.
79 Bullock (2009), p160.
80 Biggs (2012), p61.
81 Biggs (2012), p62.
82 Biggs (2012), p63.
83 Biggs (2012), p63.
84 Bullock (2009), p149.
85 Biggs (2012), pp 54–55.
86 Biggs (2012), p55.
87 Biggs (2012), p55.
88 Biggs (2012), p58.
89 Bullock (2009), p151.
90 Bullock (2009), p151.
91 Biggs (2012), p66.
92 Biggs (2012), p66.
93 Bullock (2009), p162.
94 Biggs (2012), p71.
95 Biggs (2012), p71.
96 Biggs (2012), p72.
97 Biggs (2012), p72.
98 Bullock (2009), p162.
99 Bullock (2009), p163.
100 Biggs (2012), p73.
101 Biggs (2012), p73.
102 Biggs (2012), p73.
103 Biggs (2012), p73.
104 Baudot and Dilson (1981), p74.
105 Trench (1988), p289.
106 Bullock (2009), p164.
107 Masters (1961), p319.
108 Masters (1961), p319.
109 Masters (1961), p320.
110 Bullock (2009), p164.

7 Gurkha Museum (Undated A), p5. to: Citation for Lance Corporal Tuljung
to: 7 The London Gazette, 4 October 2013, p19570

CHAPTER 7

1 Chapple (2014), p124 (gives the figure as 'approx 137,883').
2 Bullock (2009), p164.
3 Chapple (2014), p116.
4 Gurkha Museum (undated reconciliation).
5 Bullock (2009), p167.
6 Bullock (2009), p167.
7 Bullock (2009), p167.
8 Bullock (2009), p167.
9 Bullock (2009), p168.
10 Bullock (2009), p168.
11 Cross (1986), p16.
12 Bullock (2009), p169.
13 Cross (1986), p18.
14 Cross (1986), p16.
15 Cross (1986), p18.
16 Cross (1986), p18.
17 Cross (1986), p18.
18 Bullock (2014), p1.
19 Cross (1986), p20.
20 Cross (1986), p20.
21 Garran (2009), p90.
22 Cross (1986), p21.
23 Bullock (2009), p171.
24 Parker (2000), p230.
25 Allen (1991), p3.
26 Allen (1991), p3.
27 Cross (1986), p29.
28 Cross (1986), pp 29–30.
29 Cross (1986), p35.
30 Bullock (2009), p173.
31 Cross (1986), p36.
32 Cross (1986), p36.
33 Cross (1986), p38.
34 Cross (1986), p39.
35 Cross (1986), p39.
36 Cross (1986), p39.
37 Cross (1986), p39.
38 Cross (1986), p40.
39 Cross (1986), p42.
40 Cross (1986), p45.
41 Cross (1986), p45.
42 Cross (1986), p45.
43 Cross (1986), p45.
44 Cross (1986), p46.
45 Cross (1986), p48.
46 Allen (1991), p36.
47 Allen (1991), p36.
48 Allen (1991), p36.
49 Cross (1986), pp 49–51.
50 Cross (1986), pp 51–52.
51 Bullock (2009), p178.
52 Cross (1986), p70.
53 Cross (1986), p70.
55 Cross (1986), p70.
55 Cross (1986), p70.
56 Cross (1986), p71.
57 Cross (1986), p71.
58 Bullock (2009), p179.
59 Cross (1986), p83.
60 Cross (1986), p83.

CHAPTER 8

1 Chapple (2014), p172.
2 Bullock (2009), p183.
3 Allen (1991), p69.
4 Allen (1991), p68.
5 Allen (1991), p69.
6 Allen (1991), p69.
7 Bullock (2009), 183.
8 Allen (1991), p71.
9 Allen (1991), p78.
10 Allen (1991), p78.
11 Allen (1991), p79.
12 Bullock (2009), p184.
13 Bullock (2009), p184.
14 General Walker was awarded a KCB in 1968, becoming General Sir Walter Walker.
15 Bullock (2009), p184.
16 Allen (1991), p80.
17 Allen (1991), p82.
18 Allen (1991), p82.
19 Allen (1991), p82.
20 Allen (1991), p82.
21 Bullock (2009), p185.
22 Bullock (2009), p188.
23 Bullock (2009), p188.
24 Bullock (2009), p188.
25 Bullock (2009), p190.
26 Bullock (2009), p190.
27 Bullock (2009), p192.
28 Allen (1991), p95.
29 Allen (1991), p93.
30 Biggs (2012), p75.
31 Biggs (2012), p76.
32 Biggs (2012), p76.
33 Biggs (2012), p76.
34 Allen (1991), p81.
35 Bullock (2009), p197.
36 Bullock (2009), p197.
37 Chapple (2014), p191.
38 Allen (1991), p85.
39 Bullock (2009), p199.
40 Bullock (2009), p197.
41 Bullock (2009), p199.
42 Bullock (2009), p202.
43 Bullock (2009), p202.
44 Bullock (2009), pp 206–207.
45 Bullock (2009), p208.
46 Bullock (2009), p209.
47 Cross (1986), p169.
48 Cross (1986), p169 citing the citation for Lance Corporal Aimansing's Queen's Gallantry Medal.
49 Cross (1986), p169.
50 Cross (1986), p173.
51 Cross (1986), p173.
52 Bullock (2009), p210.
53 Bullock (2009), p212.
54 Cross (1986), p176.
55 Cross (1986), p178.
56 Cross (1986), p178.
57 Cross (1986), p179 (footnote).
58 Bullock (2009), p223.
59 Bullock (2009), p231.
60 Bullock (2009), p241.
61 Bullock (2009), p247.
62 Bullock (2009), p252.

CHAPTER 9

1 The QRF, or Quick Reaction Force, is a body of troops held at high readiness which can deploy at short notice to assist an outstation in the event of an incident.
2 www.Helmand.gov.af/en/page/5201 accessed on 12 February 2015.
3 Bullock (2009), p259.
4 Bullock (2009), p259.
5 Citation for Major Pitchfork's Military Cross.
6 Citation for Acting Sergeant Dipprasad Pun's Conspicuous Gallantry Cross.
7 Citation for Acting Sergeant Dipprasad Pun's Conspicuous Gallantry Cross.
8 Citation for Lance Corporal Tuljung Gurung's Military Cross.
9 14 from The Royal Gurkha Rifles and 2 from The Queen's Gurkha Engineers.

SPECIAL INTEREST SECTIONS

The Gurkha Kukri

1 Gurkha Museum (Undated A), p1.
2 Gurkha Museum (Undated A), p1.
3 Gurkha Museum (Undated A), p8.
4 Gurkha Museum (Undated A), p2.
5 Gurkha Museum (Undated A), p5.
6 Gurkha Museum (Undated A), p5.
7 Citation for Lance Corporal Tuljung Gurung's Military Cross, The London Gazette, 4 October 2013, p19570.

Hill Racing in the Brigade of Gurkhas

1 Gurkha Museum (2009), p1.
2 Gurkha Museum (2011), p20.

Bagpipes and Tartan in the Brigade of Gurkhas

1 Gurkha Museum (2011), p5.

Bands wiin the Brigade of Gurkhas

1 Chapple (2014), p190.
2 Chapple (2014), p190.

Gurkha Paratroopers

1 Loftus-Tottenham (Undated), p39.
2 Loftus-Tottenham (Undated), p38.
3 Loftus-Tottenham (Undated), p40.
4 www.G200e.com accessed on 28 February 2015.
5 Gurkha Museum (Undated B), p4.
6 Loftus-Tottenham (Undated), p41.
7 Gurkha Museum (Undated B), p4.
8 Gurkha Museum (Undated B), p4.
9 Edwards (1997), p7.
10 Chapple (2014), p185.
11 Gurkha Museum (Undated B), p4.
12 154 (Gurkha) Parachute Battalion (later 3rd Battalion, India Parachute Regiment) transferred to 77th Indian Parachute Brigade in the summer of 1945.
13 Gurkha Museum (Undated B), p4.
14 http://www.paradata.org.uk/media/3779?mediaSection=Highlights&mediaItem=3781 accessed on 1 March 2015.
15 Lowe (1990), p79.
16 Trench (1988), p289.
17 Chapple (2014), p185.
18 Chapple (2014), p185.
19 Chapple (2014), p185.
20 Chapple (2014), p191.
21 Dickens (1983), p58.
22 Dickens (1983), p58.
23 Walker (1997), p199.
24 Dickens (1983), p71.
25 Dickens (1983), p72.
26 Chapple (2014), p191.
27 Bullock (2009), p249.
28 Loftus-Tottenham (Undated), p40.

The Victoria Crosses of the Brigade of Gurkhas

1 Biggs (2012), p7.
2 Biggs (2012), p7.
3 Biggs (2012), p25.
4 www.kaiserscross.com accessed on 15 March 2015.
5 Biggs (2012), p25.
6 Biggs (2012), p25.
7 Biggs (2012), p29.
8 Biggs (2012), p36.

Select Bibliography

In writing this book, I have drawn heavily on Brigadier Christopher Bullock's definitive history of the Brigade of Gurkhas, using it as a handrail to guide me in telling the remarkable story of the last 200 years. But I have also used numerous other sources to add detail and explore events that are not included in Brigadier Bullock's book. These sources range from regimental histories, journals, diaries and webpages through to historical books, academic papers and contemporaneous magazine and newspaper articles. Inevitably, some of the 'facts' in these accounts, particularly as they relate to the early years, are contradictory and/or disputed. This selective bibliography is therefore intended to provide the source of the 'facts' as I present them in the narrative. The endnotes for each chapter explain which of the below sources support which 'fact'. Sources which provided a general understanding but not a specific 'fact' are not included.

Allen, Charles (1991), *The Savage Wars of Peace* (London: Futura)

Baudot, Marcel and Jesse Dilson (1981), *The Historical Encyclopedia of World War II* (Macmillan)

Bellamy, Chris (2011), *The Gurkhas: Special Force* (Hachette)

Biggs, Maurice (2012), *The Story of Gurkha VCs* (FireStep)

Bullock, Christopher (2009), *Britain's Gurkhas* (London: Third Millennium Publishing Limited)

Caplan, Lionel (1995), *Warrior Gentlemen: Gurkhas in the Western Imagination* (Berghahn Books)

Carey, Bertram S and H N Tuck (1896), *The Chin Hills: A history of the people, our dealings with them, their customs and manners, and a Gazetteer of their Country, Volume I.* (Rangoon)

Chapple, Field Marshal Sir John (2014) *The Lineages and Composition of Gurkha Regiments in British Service* (FireStep)

Collett, N A (2005), *The Butcher of Amritsar: General Reginald Dyer* (A&C Black)

Cross, J P (1986), *In Gurkha Company: the British Army Gurkhas, 1948 to the Present* (Arms and Armour Press)

Dickens, Peter (1983), *SAS: The Jungle Frontier* (London: Book Club Associates)

Ferguson, Niall (2003), *Empire: How Britain Made the Modern World* (Allen Lane/ Penguin Press)

Garran, Robin (2009), 'When Gurkhas Were Gunners', *7th Duke of Edinburgh's Own Gurkhas Regimental Association Journal*, Number 15, pp 90–92

Gould, Tony (1999), *Imperial Warriors: Britain and the Gurkhas* (London: Granta Books)

Gurkha Museum (2009), *The Story of Gurkha Hill Racing* (Gurkha Museum)

Gurkha Museum (2011), *The Scottish Connection* (Gurkha Museum)

Gurkha Museum (Undated A), *Brief – The History of the Kukri* (Gurkha Museum)

Gurkha Museum (Undated B), *The Gurkha Parachutist* (Gurkha Museum)

Hill, James and Horace Hayman Wilson (1858), *The History of British India From 1805 to 1835, Volume 9* (London: James Madden)

Johnson, Rob (2011), *The Afghan Way of War: How and Why They Fight* (New York: Oxford University Press)

Knight, Paul (2013), *The British Army in Mesopotamia, 1914–1918* (McFarland)

Loftus-Tottenham, Major General F J (Undated), *Walkabouts and Laughabouts in the Raj* (2nd Gurkhas)

Low, Charles Rathbone (1883), *Major General Sir Frederick S Roberts: A Memoir* (London: W H Allen & Co)

Lowe, Henry (1990), 'The Battle of Sangshak', *The RUSI Journal*, Volume 135, Issue 1, pp79-80.

Lunt, James (1994), *Jai Sixth! 6th Queen Elizabeth's Own Gurkha Rifles 1817–1994* (Pen and Sword)

Marshall Cavendish (2002), *History of World War I, Volume 1* (New York: Marshall Cavendish)

Marshall Cavendish (2002b), *History of World War I, Volume 2* (New York: Marshall Cavendish)

Marshall, H E (2006), *Our Empire Story* (Yesterday's Classics – Reprint of 1908 Classic)

Masters, John (1961), *The Road Past Mandalay*

Meyers, Jeffrey (2004), 'T. E. Lawrence and the Character of the Arabs', *The Virginia Quarterly Review* 80, Number 4

Morris, James (1979), *Heaven's Command: An Imperial Progress* (Penguin)

Mortlock, Michael J (2010), *The Egyptian Expeditionary Force in World War 1: A History of the British-Led Campaigns in Egypt, Palestine and Syria* (McFarland)

Murland, H F (2012), *Baillie-Ki-Paltan: Being a History of the 2nd Battalion, Madras Pioneers 1759–1930* (Andrews)

Murphy, David (2012), *Lawrence of Arabia* (Osprey)

New Zealand Herald (1891), *The Manipur Rising*, Volume XXVIII, Issue 8,573 dated 22 May 1891.

Palace, Wendy (2004), *The British Empire and Tibet, 1900–1922* (New York: Routledge)

Parker, John (2000), *The Gurkhas* (Headline Book Publishing – Paperback Edition)

Pemble, John (2009), 'Forgetting and Remembering Britain's Gurkha War', *Asian Affairs*, Volume 40, Issue 3, pp 361–376

Prinsep, Henry T (1820), *A Narrative of the Political and Military Transactions of British India under the Administration of the Marquess of Hastings, 1813–1818* (London: John Murray)

Raugh, Harold E (2004), *The Victorians at War, 1815–1914: An Encyclopaedia of British Military History* (ABC Clio)

Read, Anthony and David Fisher (1998), *The Proudest Day: India's Long Road to Independence* (London: Random House)

Ross (2005), *Atlas of 20th Century Warfare* (Arcturus Publishing Limited)

Shore, F J (1836), 'Report on the Dehra Doon 1827–28', *Calcutta Monthly Journal and General Register of Occurrences*, June 1836

Slim, Field Marshal Viscount (1956), *Defeat into Victory* (Cassell and Co)

The Argus (Melbourne, Victoria) (1891), Monday 30 November 1891.

Thorn, Sir William (1816), *A Memoir of Major General Sir Rollo Gillespie* (T Egerton)

Trench, Charles Chenevix (1988), *The Indian Army and the King's Enemies 1900–1947* (Thames and Hudson)

Walker, General Sir Walter (1997), *Fighting On*

Wilson, H H (1858), *The History of British India from 1805–1835, Volume II* (London: James Madden)

Woodburn, Bill (2012), 'The First Anglo-Sikh War', *Asian Affairs*, Volume 43, Issue 1, pp 146–147

Acknowledgements

This book would not have been possible without a huge amount of support, given with constant good humour and at no cost, from a whole range of people and organisations. Particular thanks are due to Gavin Edgerley-Harris, the Curator of the Gurkha Museum, and to Brigadier (retired) Christopher Bullock OBE MC, author of the definitive history of the Brigade of Gurkhas. These two remarkable people have been unstinting in their support over many months. Not only have they checked my work for historical accuracy but they have also worked hard to improve the quality of the text and the relevance of the imagery. The final product owes a great deal to them both.

I am also very grateful to the staff at Headquarters Brigade of Gurkhas who have spent countless hours finding imagery and sorting out the authority for me to use pictures of serving soldiers. Charles Heath-Saunders at Army Headquarters helped me obtain Ministry of Defence permission to publish the book and Major Tom Usher, now retired from The Royal Gurkha Rifles, helped me collate recent images of the serving Brigade from Afghanistan. Major (retired) Bruce McKay MBE, the Regimental Secretary of The Royal Gurkha Rifles, helped with imagery and provided administrative support whilst Colonel (retired) William Shuttlewood OBE, the Director of the Gurkha Welfare Trust, provided constant support throughout the production process. My publisher, Ryan Gearing, provided guidance, wisdom and encouragement, patiently helping me deliver something worthwhile despite the many delays that my operational deployment caused. Nick and George Newton, the designers who assembled the book from its many constituent parts, have been quite exceptional in bringing the text and imagery to life. I am grateful to the book's sponsors, and to Ian Bowles at Allocate Software in particular, for their financial support without which we would have been unable to produce such a high quality product. Special thanks are also due to His Royal Highness, The Prince of Wales and to Miss Joanna Lumley OBE for supporting the book and for agreeing to write the foreword and the introduction respectively.

My family, and my wife Laura in particular, deserve special mention. It has taken me nearly two years to produce this book. Throughout, my family has been hugely supportive despite the many holidays, weekends and evenings that I have spent working on it rather than spending time with them. Laura has been both a tower of strength and a constant source of good ideas. Without her good humoured encouragement, this book would not have been produced.

Finally, my heartfelt thanks go to the officers and soldiers of Britain's Brigade of Gurkhas, past, present and future. I have tried to describe the last 200 years of their Service to the Crown. Their many achievements, which have come at a very real human cost, have made writing this book both an honour and a pleasure. My only hope is that I have done them justice.

Index

Post-nominal decorations are not shown against an individual's name unless they relate to the reason for inclusion in the index (such as an award for gallantry)